COLUMBIA REVIEW

HIGH-YIELD ORGANIC CHEMISTRY

Stephen D. Bresnick, M.D.

President and Director
Columbia Review, Inc.
San Francisco, California

Williams & Wilkins
A WAVERLY COMPANY

BALTIMORE • PHILADELPHIA • LONDON • PARIS • BANGKOK
BUENOS AIRES • HONG KONG • MUNICH • SYDNEY • TOKYO • WROCLAW
1996

Editor: Elizabeth A. Nieginski
Managing Editor: Alethea H. Elkins
Production Coordinator: Danielle Santucci
Copy Editor: Steve Jensen
Designer: Ashley Pound Design
Typesetter: Maryland Composition Co., Inc.
Printer: Port City Press
Binder: Port City Press

351 West Camden Street
Baltimore, Maryland 21201-2436 USA

Rose Tree Corporate Center
1400 North Providence Road
Building II, Suite 5025
Media, Pennsylvania 19063-2043 USA

Printed in the United States of America

Library of Congress Cataloging in Publication Data

The Publishers have made every effort to trace the copyright holders for borrowed material. If they have inadvertently overlooked any, they will be pleased to make the necessary arrangements at the first opportunity.

96 97 98 99
1 2 3 4 5 6 7 8 9 10

To purchase additional copies of this book, call our Customer Service Department at (800) 638-0672 or fax orders to (800) 447-8438. For other book services, including chapter reprints and large quantity sales, ask for the Special Sales Department.

Canadian customers should call (800) 268-4178, or fax (905) 470-6780. For all other calls originating outside the United States, please call (410) 528-4223 or fax us at (410) 528-8550.

Visit Williams & Wilkins on the Internet at http://www.wwilkins.com or contact our Customer Service Department via e-mail at custserv@wwilkins.com. Williams & Wilkins customer service representatives are available from 8:45 am to 6:00 pm, EST, Monday through Friday, for either telephone or Internet access.

RECYCLED

Contents

SECTION II

The Chemistry of Oxygen Containing Organic Compounds

Organic Molecules of Biologic Importance and Separation/Purification of Organic Compounds

vii

Review Questions

Preface

High-Yield Organic Chemistry is an easy-to-read, efficient, and high-quality review book for first-year, college-level organic chemistry. The book focuses on a conceptual review of core organic chemistry topics and for its size covers an amazing amount of material. For mastery of review material, over 340 practice questions with solutions are provided. The book is designed for all college students or others wishing to understand and review the major concepts of organic chemistry. Students who are pre-health, chemistry, or non-science majors will benefit from this book.

High-Yield Organic Chemistry is one of four books in the *High-Yield Undergraduate Science Review Series* by Williams & Wilkins. The series also contains *High-Yield General Chemistry, High-Yield Physics,* and *High-Yield Biology.* This series has been designed to make these four important college sciences easier to understand and master. All the High-Yield books contain a science review, many examples and sample problems, and several hundred practice questions with answers and explanations.

The author of this series, Dr. Stephen Bresnick, is an expert in helping students understand, review, and retain basic college science material. Dr. Bresnick understands that many students work their way through college courses without really comprehending the material they are supposed to be learning. He has designed these four books to help students **understand science better** and **improve their course grades.** In addition, the series has been designed to help students prepare for **post-graduate and pre-professional tests,** such as the GRE, MCAT, DAT, PCAT, VET, OAT, and other tests. Dr. Bresnick is a physician and educator who both teaches and writes science review material for college students. He is currently Director of **Columbia Review,** a national test-preparation company specializing in science and English review for pre-medical school students.

Organization

There are three sections in this book. Each section corresponds to the specific topics that most college students study in organic chemistry courses. The topic review emphasizes conceptual learning and provides numerous sample problems and examples. Each section of the book is followed by several sets of review questions and their solutions.

Acknowledgments

The author wishes to thank Dr. William Bresnick and Dr. Nori Kawahata for their contributions. In addition, many thanks to the staff of Williams & Wilkins for their dedication in creating a great high-yield review book for organic chemistry. I especially wish to thank Lee Elkins, Danielle Santucci, Elizabeth Nieginski, Jane Velker, Tim Satterfield, and Kevin Thibodeau for their expertise and assistance with this important project.

General Concepts and Hydrocarbon Chemistry

org

anic

chemi

.....

General Concepts

concept

I. Basic Introduction

Success in organic chemistry requires a thorough knowledge and understanding of the basics. These review notes show an assortment of reactions that give a general overview of organic chemistry. **Do not just memorize reactions!** Instead, familiarize yourself with how and why the reactions occur. Use the principles and concepts presented in the Organic Chemistry Review Notes and subsequent practice to gain an understanding of the types of processes that molecules undergo as they try to attain more favorable energy states. The key is to understand trends and principles.

The appendix to the Organic Chemistry Review Notes provides a summary of the rules for nomenclature in organic chemistry. As you review each functional group, be sure to flip back to this appendix to review the nomenclature for that group.

So what are the basics?

1. Atoms are mostly fluffy electron clouds (atomic orbitals). What differentiates them is the number of electrons in the outer shell and how far away this shell is from the nucleus. These outer shell electrons determine physical properties and reactivity.
2. Molecules are formed when two atoms join their atomic orbitals to form a larger electron cloud [molecular orbital (MO)]. The sharing of electrons in an MO constitutes a covalent bond. Electrons can move within an MO and they gather around the more electronegative atom of the bond. This focus of electrons in particular regions of a molecule affects the physical properties and chemical reactivities of the molecules.
3. In most cases, bonds form between species that are electron deficient (**electrophiles**) and those that are electron rich (**nucleophiles**). Thus, it is important to be able to recognize which types of atoms and molecules are nucleophiles and which are electrophiles.
4. When analyzing chemical reactions, remember that a reaction is a collision between two molecules. Organic reactions (i.e., bonds breaking and forming) occur because they involve energy-releasing processes or because enough energy is applied to force the reaction. In general, everything strives toward lower energy.
5. Lower-energy species are more stable and less reactive, whereas higher-energy species are less stable and more reactive.

II. Electronegativity and Bond Formation

A. IMPORTANCE OF ELECTRONEGATIVITY AND VALENCE ELECTRONS IN BOND FORMATION

A great way to remember the electronegativity order of key elements, going from the highest electronegativity to the lowest, is to learn the expression, **FONClBrISCH (pronounced "fawn-**

cul-brish"). This expression will help you to remember that the key elements (fluorine, oxygen, nitrogen, chlorine, bromine, iodine, sulfur, carbon, and hydrogen) descend predictably in electronegativity strength compared to one another. **Electronegativity is defined as the ability of an atom to attract electrons.**

From this order, recognize that **electronegativity increases as you move to the right on the periodic table.** Halogens (group VII), with seven valence electrons, need only one more electron to attain the energetically favorable complete outer-shell octet. Thus, they have a relatively high affinity for electrons. After gaining one electron, group VI elements still need another electron to reach a complete octet; thus, their affinity for an electron is not as great as that of the halogens. Group V atoms have an even lower affinity for an electron.

Electronegativity decreases as you move down the periodic table. Concurrently, the atomic radii of the elements increase greatly. An electron added to the outer shell of a relatively large atom such as iodine does not feel as strong a nuclear charge (because of **shielding** by the electrons in energy levels below) as an electron added to fluorine. Also, the outer-shell electrons of a large atom such as iodine are not held as tightly as those of smaller atoms because of their greater distance from the positively charged nucleus.

As you move to the left on the periodic table, the elements are increasingly willing to give up electrons. By releasing a few electron(s), the next lower energy level that contains a full octet becomes the shell. For example, in group IA, Na^+ is Na that has given up one electron to expose a full outer shell of electrons.

All atoms not in this list have electronegativities lower than hydrogen. All members of FONClBrISCH (except H) are on the right side of the periodic table. They need to gain three or fewer electrons to obtain a complete octet of electrons in their outer shell.

The number of valence electrons determines how many bonds are formed. Halogens (group VII) have seven electrons in their outer shell and generally form one bond to establish a complete octet. Atoms in group VI (oxygen) have six valence electrons, so they generally form two bonds to complete an octet. Nitrogen and other group V elements form three bonds to complete their octet. Carbon, one column to the left on the periodic table, has four valence electrons and forms four bonds.

B.   TYPES OF BONDS

1.   IONIC BONDS

Ionic bonds occur between two atoms of different electronegativities. The electronegative atom takes an electron from another atom that possesses a low ionization potential (willing to give up an electron to obtain a completely filled outer shell of electrons).

Typical ionic bonds involve halogens and metals (e.g., NaCl, KI). **These bonds are strong.** Furthermore, crystal lattices form, involving a highly ordered system with extensive intermolecular bonding. The intermolecular bonds give these compounds high melting and boiling points. To solvate these molecules, the intermolecular interactions of the solid must be replaced by interactions with the solvent. In other words, polar solvents must be used to solvate these highly polar molecules. Ionic bonds also occur between charged clusters of atoms ($NH_4^+OH^-$, $NH_4^+CH_3COO^-$).

2.   COVALENT BONDS

Covalent bonds involve the sharing of electrons, but it is not a "give-and-take" relationship as with ionic bonds. They occur in diatomic molecules, such as O_2 or Cl_2, and between atoms with similar electronegativities. **Polar covalent bonds** involve atoms that have different

4

electronegative strengths, but not to the point where an ionic bond forms. The movement of electrons toward the more electronegative atom results in a partial positive charge on one end of the bond (lower electronegativity end) and a partial negative charge on the other end (higher electronegative end).

3.   HYDROGEN BONDS (H-BONDS)

Hydrogen bonds are relatively weak intermolecular interactions in which a slightly acidic (partially + or δ^+) hydrogen forms a weak dipole interaction with a neighboring basic (partially − or δ^-) atom. Look for hydrogen atoms attached to N, O, and F. These atoms tend to be good donors of hydrogen toward H-bonds. Good acceptors are electronegative atoms with lone pair electrons, such as O, N, and F. Hydrogen bonds account for many physical properties of compounds in organic chemistry.

C.   ATOMIC, MOLECULAR, AND HYBRID ORBITALS

1.   GENERAL IDEAS

The **atomic orbitals (AO)** are theoretic regions around the nucleus where the probability of finding an electron is high. Each energy level contains its own AO.

The **covalent bond** involves a sharing of electrons. **This bond between two atoms is an overlap of two atomic orbitals.** The overlap of two AO forms two **MO**—one bonding MO (lower energy) and one antibonding MO (higher energy). Electrons fill the lower-energy bonding MO first, and then they fill the higher-energy antibonding MO. If the number of electrons in the bonding orbitals is the same as that in the antibonding orbitals, no bond will form between the two atoms.

The **hybridization** of AO involves a mixing of AO on an atom to create a hybrid AO. These new AO allow for greater overlap when forming MO, leading to a stronger (lower-energy) bond. The hybridization determines the shape of the molecule.

2.   ATOMIC ORBITALS

Atomic orbitals describe a cloud in which electrons of a particular energy level are likely to be found about the nucleus of an atom. These orbitals have specific shapes, depending on their energy.

For organic chemistry, it is important to know the following:

s orbital: Spherically symmetric about the nucleus
p orbital: Dumbbell shaped with nucleus at the center

With the exception of the first energy level, which contains only the s orbital, each principal energy level has one s orbital and three p orbitals arranged 90° from each other (Figure 1-1).

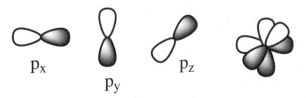

Figure 1-1.   The arrangement of the p orbitals.

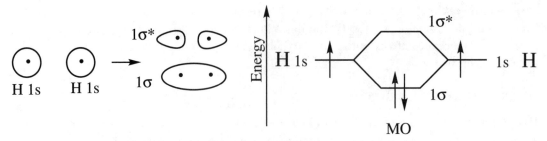

Figure 1-2. Molecular orbital diagram for H_2.

Each orbital can hold two electrons. Remember that the filling order is $1s^2$, $2s^2$, $2p^6$, $3s^2$, $3p^6$, $4s^2$, $3d^{10}$, $4p^6$, $5s^2$, $5p^6$, and so on, filled from lowest to highest energy.

3.   MOLECULAR ORBITALS

Molecular orbitals, one bonding and the other antibonding, are created when atomic orbitals from two atoms combine and form a single electron cloud. The two electrons in this MO are shared by the contributing atoms. Each MO can hold two electrons. Two simple examples follow.

Example 1-1: H_2 (Figure 1-2)

a.   Two 1s orbitals combine to form 1σ (bonding) and $1\sigma^*$ (antibonding) MO.
b. σ represents an MO formed from a direct **end-to-end overlap of AO.**
c. σ^* represents a molecular orbital with **no atomic orbital overlap.**
d.   Two electrons (one from each H) fill the 1σ MO, and a bond forms.

Example 1-2: He_2 (Figure 1-3)

a.   Four electrons (two from each He) fill the 1σ and $1\sigma^*$ MO.
b.   Because both the higher-energy $1\sigma^*$ MO and the 1σ MO are occupied, a bond does not form between two He atoms. This type of orbital filling occurs for all noble gases and is the reason why they do not exist as diatomic species. It is energetically preferable for electrons to remain in the lower-energy AO. A bond will form only when electrons fill more "bonding" MO than antibonding MO.

This diagrammatic method is an easy way to conceptualize simple molecules. Such diagrams become exponentially more difficult as more electrons and atoms are added to the picture, but they give you an idea of what to think about as you draw a line on paper to designate a bond.

Remember that bonds form because it is energetically more favorable for the atoms to be bonded rather than unbound. In Figures 1-2 and 1-3, note that the energy of the 1σ MO is lower

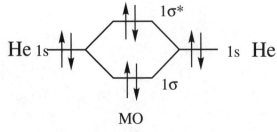

Figure 1-3. Molecular orbital diagram for He_2.

6

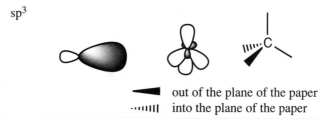

sp³

◀ out of the plane of the paper
·····ııııı into the plane of the paper

Figure 1-4. The sp³ hybrid. The solid arrow, out of the plane of the paper; the dashed arrow, into the plane of the paper.

than that of the 1s AO. Thus, the electrons of H_2 are in a lower energy state than they would be individually. Conversely, electrons occupying the He 1s AO are lower in energy than the electrons in the $1\sigma^*$ MO, and they prefer to stay that way. Thus, He is not a diatomic gas. **Often when you break a bond and form another in an organic reaction, you are removing the electrons from a higher-energy MO and placing them into a lower-energy MO.**

4. HYBRID ORBITALS

Hybridized atomic orbitals are created when atomic orbitals on an atom mix together. Carbon contains four electrons in its valence shell ($2s^2$, $2p^2$). The s and p orbitals can be mixed to form the sp^3, sp^2, or sp hybrid orbitals. The types of hybrids and the formation of double and triple bonds are shown in Figures 1-4 to 1-8.

a. sp³ (Figure 1-4)

The sp^3 hybrid orbitals can be thought of as one s and three p orbitals mixing to form four sp^3 orbitals. In Figure 1-4, a single sp^3 hybrid orbital is shown at left and the orientation of all four of the sp^3 hybrids and their geometry are shown at right.

   The four sp^3 hybrids, each containing one electron, can form four σ (direct overlap) bonds. The hybrid orbitals are arranged **tetrahedrally** about the nucleus to maximize the distance between the electrons in the four σ-bonds. Thus, the bonds of an sp^3 hybridized carbon have a bond angle $\approx 109.5°$. These orbitals are longer than s orbitals and more ovoid than p orbitals. This shape allows for a greater overlap volume with AO of other atoms when forming MO. Larger overlap results in lower energy and more stable MO.

b. sp² (Figure 1-5)

One s and two p orbitals mix to form three sp^2 orbitals, leaving one unhybridized p orbital. The three sp^2 orbitals are arranged on the same plane (120° apart) and are perpendicular to the axis of the p orbital. Each of the sp^2 orbitals is equal in energy and contains one electron. The unhybridized p orbital is slightly higher in energy and also contains one electron.

   The sp^2 orbitals can form a σ-bond (direct end-to-end overlap of AO) and the **p orbital can form a π-bond with a properly aligned p orbital, resulting in a double bond** (Figure 1-6).

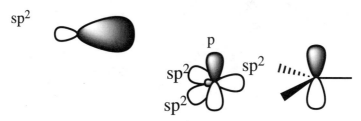

Figure 1-5. The sp² hybrid.

Figure 1-6. The double bond.

π-Bonds, unlike σ-bonds, are sideways overlaps of p orbitals. The overlap is not as great as with end-to-end overlaps, which occur for σ-bonds. Thus, the π-MO is higher in energy compared with the σ-MO of a double bond. In a σ-bond, the electrons are found mainly between the two nuclei. For a π-bond, the electrons are mostly found above and below the bond and are more accessible for electrophilic species.

c. sp (Figure 1-7)

One s and one p orbital mix to form two sp orbitals, leaving two unhybridized p orbitals. The two sp orbitals are arranged 180° apart and are perpendicular to the p orbitals (which are perpendicular to each other). **For carbon,** each of the two hybridized orbitals and the two unhybridized p orbitals contains one electron.

The sp orbitals can each form a σ-bond and the two **p orbitals can form π-bonds with two properly aligned p orbitals, resulting in a triple bond** (Figure 1-8). Think of the π-electrons surrounding the two nuclei in a barrel-shaped cloud of electrons.

Note: Atoms connected with just σ-bonds can be rotated about the bond. Those atoms connected by σ- and π-bonds, however, require more energy, because to rotate about a double bond, the π-bond must be broken. Thus, compounds with multiple bonds are more rigid than those with just single bonds.

D. HYBRIDIZATION, MOLECULAR SHAPE, AND POLARITY

The type of hybridization determines the shape of the molecule, which in turn determines the polarity of the molecule. **Hybridization occurs with atoms other than carbon, including nitrogen** [sp^3 in ammonia (NH_3)] and **oxygen** (sp^3 in water and ethers). **Boron** is sp^2 hybridized, with an electron in each hybridized orbital and an empty p orbital. The resulting trigonal planar shape (Figure 1-9) of boron results from the hybridized orbitals and the empty p orbital(s).

1. POLARITY

Recall that polar covalent bonds are dipoles consisting of an electron-rich region and an electron-deficient region. The net polarity of a molecule is the vector sum of its component dipoles.

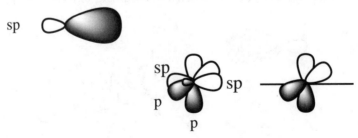

Figure 1-7. The sp hybrid.

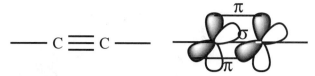

Figure 1-8. The triple bond.

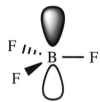

Figure 1-9. The sp² hybridization of BF₃.

Dipole moments measure the polarity of molecules as a whole. For example, Figures 1-10 and 1-11 show different types of hybridized molecules with or without dipoles. Note that the net dipole moment is derived by the vector sum of the bond moments of the molecules. Thus, CCl_4 and CO_2 have no net dipole moment, whereas the other molecules have a net dipole moment.

Why is polarity important? The net polarity of a molecule affects the interactions of that molecule with other molecules (of its own kind or different ones). Such physical properties as melting and boiling point, density, and solubility are affected by the degree of intermolecular interaction.

2.   SOLUBILITY AND SOLVATION

The phrase "like dissolves like" means that only polar solvents can solvate polar compounds. Solvation involves the breaking of intermolecular interactions of the solute and replacing them with

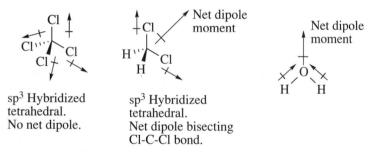

sp³ Hybridized tetrahedral.
No net dipole.

sp³ Hybridized tetrahedral.
Net dipole bisecting Cl-C-Cl bond.

Net dipole moment

Net dipole moment

Figure 1-10. Examples of sp³ hybridized molecules with or without dipoles (CCl_4, CH_2Cl_2, and H_2O).

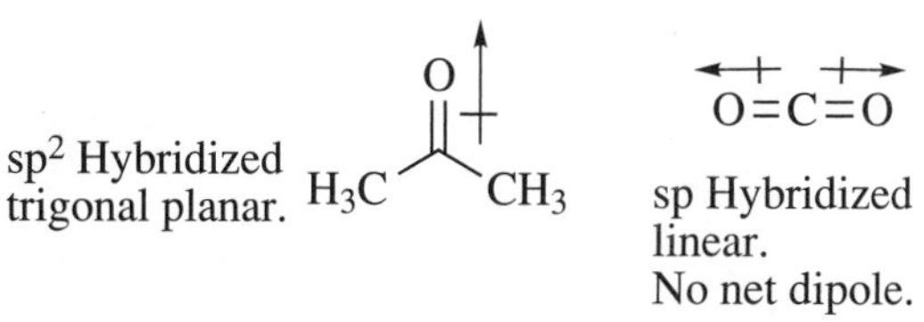

sp² Hybridized trigonal planar.

sp Hybridized linear.
No net dipole.

$O=C=O$

Figure 1-11. Examples of sp² and sp hybridized molecules with or without dipoles (acetone, CO_2).

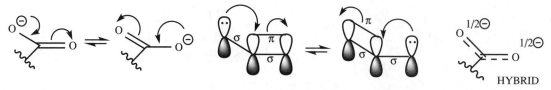

Figure 1-12. The resonance forms and resonance hybrid for carboxylate. *Arrows* represent movement of two electrons.

interactions of the solute with the solvent. Nonpolar solvents cannot interact with polar solutes and are thus unable to dissolve polar compounds. Nonpolar solids or liquids do not dissolve in polar solvents. The new interactions between the nonpolar solute and the polar solvent are not strong enough to replace the dipole–dipole interactions stabilizing the solvent molecules [e.g., oil (nonpolar) and H_2O do not mix]. Nonpolar compounds, however, dissolve in nonpolar solvents because the molecules of nonpolar solvents are not held together by strong intermolecular interactions.

E.   RESONANCE FORMS AND DELOCALIZATION

A problem associated with Lewis structures is that they tend to position electrons in a single location (in a bond between two atoms). Two or more equivalent Lewis structures that differ only in the position of the electrons are referred to as resonance structures, or forms. An example of these forms is shown on the left in Figure 1-12. Orbital details of a pair of resonance forms are shown in the center of the figure.

The **resonance hybrid** is a structure that takes into account the individual resonance forms of a compound, and is **more stable** than any of the individual resonance forms that contribute to its formation. Resonance structures are therefore contributors to the overall resonance hybrid of a molecule. **The more stable a particular resonance form, the more it contributes to the overall resonance hybrid.** This fact indicates that electrons in a π-bond are not always located between the two atoms of the original bond. Rather, they are spread out or delocalized, which enhances stability. Note: Only π-bonded electrons or free electrons in a p orbital can delocalize in a molecule.

Neutral species are also stabilized through resonance delocalization of electrons. Benzene is highly stabilized because of multiple resonance forms; two resonance structures and the resonance hybrid are shown in Figure 1-13. Note that the **resonance hybrid for benzene suggests a single π-MO spread over the six atoms of the ring.**

When drawing resonance forms of a molecule, remember to:

1.   Move only the electrons. The positions of the atoms must be preserved.
2.   Maintain proper Lewis structures (i.e., no five-bonded carbons, and the like).
3.   Conserve the charge and the number of unpaired electrons.

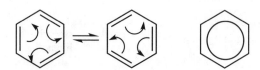

Figure 1-13.   Resonance structures and resonance hybrid for benzene.

10

More stable

Figure 1-14. Electron movements associated with resonance structures.

Figure 1-14 illustrates how to move electrons to generate a resonance structure.

To estimate the relative stability of resonance structures, look at the overall **viability** of the forms. In general, **forms with more covalent bonds are more stable.** Resonance forms in which charges are located on atoms that stabilize them (i.e., negative charges on electronegative atoms) are also more stable. Finally, forms with complete octets and fewer charges are also more stable.

III. Acid–Base Theory

A. GENERAL IDEAS

1. BRØNSTED-LOWRY DEFINITION OF ACIDS AND BASES

A Brønsted-Lowry acid is any molecule or ion that can donate a proton.
A Brønsted-Lowry base is any molecule or ion that can accept a proton.

$$\underset{\text{acid 1}}{HCl} + \underset{\text{base 2}}{H_2O} \rightleftharpoons \underset{\text{acid 2}}{H_3O^+} + \underset{\text{base 1}}{Cl^-}$$

Consider the dissociation of HCl in water: Using the Brønsted-Lowry (BL) definitions, note that this reaction includes two acids, HCl and hydronium ion, and two bases, water and chloride ion. All BL acid–base reactions involve two conjugate acid–base pairs in equilibrium. In evaluating this reaction, note that HCl acts as a BL acid by donating a proton to water. The water molecule acts as a BL base by accepting the proton. The hydronium ion can act as a proton donor to the chloride ion, and thus act as a BL acid. The chloride ion can act as a base by accepting a proton from the hydronium ion. Remember that these reactions are in equilibrium. Note also how each BL acid becomes a conjugate base, and vice versa.

2. LEWIS DEFINITION OF ACIDS AND BASES

A Lewis acid is any species that can accept an electron pair.
A Lewis base is any species that can donate an electron pair.

A Lewis acid–base reaction creates a **coordinate covalent bond** in the reaction product. The Lewis system is useful for classifying reactions that occur in solvents other than water or in the complete absence of a solvent. In the Lewis system, all cations are acids and all anions are bases.

Consider the compound BF_3 (see Figure 1-9). The boron atom shares six electrons with the three fluorine atoms and has a vacant 2p orbital that can accept a pair of electrons. Thus, BF_3 can act as a Lewis acid and accept a pair of electrons from a Lewis base. Ammonia (NH_3) is a great example of a Lewis base. The lone pair of electrons on the nitrogen of ammonia can be donated to a suitable Lewis acid, such as BF_3, to create a coordinate covalent bond.

B.   THE DISSOCIATION EQUILIBRIUM

Acids, according to Brønsted, are compounds that donate protons (protonate bases). To donate protons, the equilibrium must shift to the right. Look at the following dissociation equation:

$$HA + H_2O \rightleftharpoons H_3O^+ + A^-$$

$$K_a = [H_3O^+][A^-]/[HA] = (prod / react) \text{ or } (dissociated / protonated)$$

If the equilibrium shifts to the right, the K_a should increase. In other words, strong acids are characterized by large K_a (small pK_a) values. It follows then that weak acids are characterized by small K_a (large pK_a) values.

C.   ACIDITY TRENDS

The more stable the conjugate base (A^-), the stronger the acid (HA). If A^- is unstable (more reactive and basic), the dissociation equilibrium will shift to the left, resulting in a weaker acid.

D.   HOW TO STABILIZE THE CONJUGATE BASE (A^-)

1.   INCREASE ELECTRONEGATIVITY

Increasing the electronegativity of A^- increases acidity. More electronegative conjugate bases can better hold on to the extra electron after the proton is released from the acid.

$$HF > H_2O > NH_3 > CH_4 \text{ (stronger to weaker acid)}$$
$$pK_a \quad 3.2 \quad\quad 16 \quad\quad 38 \quad\quad 48$$

Another way of thinking is that the H-F bond is more polarized than the C-H bond because of the higher electronegativity difference. Thus, it is easier for H-F to lose an H^+.

2.   SPREAD OUT THE CHARGE

a.   Spread Over a Large Volume

This factor overrides the electronegativity trend. A charge in a small volume is more unstable because of the repulsion from nearby electrons. The resulting reactive conjugate base is more likely to shift the dissociation equilibrium to the left.

$$HI > HBr > HCl > HF \text{ (stronger to weaker acid)}$$

Iodine (I) is the least electronegative halogen, but the anion I^- is the largest, and it can disperse the negative charge over a large volume (compared with F^-), making it less apt to reverse the deprotonation reaction. Remember: **spread the charge, increase stability.**

b.   Resonance Effect

Conjugate bases with multiple resonance forms are more stable than those without, and thus, the acids are stronger. The conjugate base of acetic acid has greater charge delocalization (see carboxylate delocalization, Figure 1-12) than the conjugate of methanol (CH_3O^-), resulting in a more stable A^- for acetic acid. Again, spreading the charge stabilizes A^-.

$$CH_3COOH > CH_3OH$$

pK$_a$	4.7	16

c.   Resonance Combined with Inductive Effects

In fluoroethane (CH_3CH_2F), the presence of the electronegative (electron-attracting) fluorine places a partial positive charge (δ^+) on the CH_2 to which it is bound. This carbon, in turn, pulls electron density toward itself from the C-C bond, which places a δ^+ on the carbon of the CH_3 group. This action is called the **inductive effect.** The effect decreases with increasing distance from the electronegative substituent.

The inductive effect often increases acidity. Consider acetic acid.

$$Cl—CH_2—COOH > CH_3COOH$$

pK$_a$	2.86	4.76

Why does adding the chloride atom increase the acidity (decrease pK$_a$) of acetic acid? When an electronegative atom is placed near the functional acid group of acetic acid, the electronegativity of Cl polarizes the C-Cl bond and, through inductive effects, further polarizes the O-H bond, making the proton more δ^+. This action weakens the O-H bond and the proton is released more easily.

The inductive effect can also be thought of as stabilizing the conjugate base by helping to spread out the negative charge (Cl pulls electron density toward itself).

E.   A FEW LAST WORDS

Weaker acids have stronger conjugate bases. This statement makes sense, because weaker acids have less stable (more reactive or stronger) conjugate bases (A^-).

If HA and A^- represent a conjugate acid–base pair, then K$_a$ of HA = $[H_3O^+][A^-]/[HA]$ and K$_b$ of A^- = $[HA][OH^-]/[A^-]$. Thus, $K_a^{HA}K_b^{A^-} = [H_3O^+][OH^-] = K_w$, a constant. This formula is a general truism for any **conjugate** pair and quantifies the nonsymmetric relationship between the relative strengths of the acid and its conjugate base (i.e., **the stronger the acid, the weaker the conjugate base**).

Reactions of acids with bases: $HA + B \rightleftharpoons BH^+ + A^-$

The equilibrium favors the side with the weaker acid–base pair. Strong bases are needed to deprotonate weak acids and strong acids are needed to protonate weak bases. In the preceding example, B must want the proton more than A^-. The weaker an acid HA, the stronger a base A^-. Thus, B must be a strong base to hold on to the proton from a weak acid.

The Henderson-Hasselbalch equation

$$pH = pK_a + \log [A^-]/[HA]$$

states that when the pH of a solution is equal to the pK$_a$ of a dissolved acid, half of the acid will be protonated and the other half will be deprotonated. (See also Chapter 10.)

Energy and Kinetics

Reactions sometimes occur spontaneously (in organic chemistry, bonds break and form) because the products are more stable than the reactants. In other cases, even though the products are energetically more stable, application of energy is still needed because of the initial energy barrier that must be surpassed to move the reaction forward. We use the energy versus reaction coordinate to depict (in two dimensions) the energy of a species as it goes from one form to another. Remember that a species that is lower in energy is more stable and less reactive. **(The energy referred to is the Gibb's free energy, G.)**

A. ENTHALPY

The change in enthalpy of a reaction ($\Delta H°$; in which ° denotes standard conditions, 1 atm for a gas, 1 M for solution) measures the difference in the relative enthalpies (heat content or potential energy) of the reactants and products. The reactions are as follows:

Exothermic if $\Delta H° < 0$, indicating that heat is liberated during the course of the reaction.
Endothermic if $\Delta H° > 0$, indicating that heat is absorbed during the course of the reaction.

The energy diagrams in Figure 1-15 show the two processes; it is assumed that ΔG and ΔH have the same sign. Note that for each process, reactants pass through a transition state, have an associated activation energy, and form products.

B. ACTIVATION ENERGY

The **activation energy** barrier must be surpassed for a reaction to occur. The height of this barrier determines the rate at which the reaction occurs. Enough energy must be available to push a reaction forward to the **transition state** (point B of the energy diagrams). The transition state, where bonds are being partially formed or broken, represents an intermediate between reactants and products. Once the reaction reaches this point, it can roll either way (forward or back) depending on the conditions and the equilibrium. **Catalysts** accelerate a reaction by decreasing the energy difference between the starting material and the transition state, thus decreasing the activation energy.

C. GIBB'S FREE ENERGY: $\Delta G°$ STANDARD FREE ENERGY CHANGE

The standard free energy change ($\Delta G°$) measures the driving force or the push of a reaction. When $\Delta G° < 0$, the reaction is considered spontaneous. The more negative the $\Delta G°$ of a reac-

Energy Diagrams

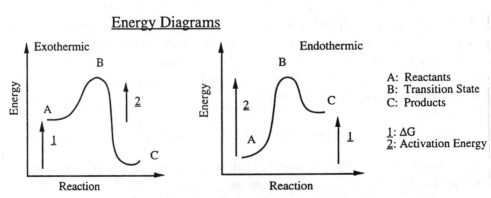

Figure I-15.   Exothermic and endothermic reactions.

tion, the greater the tendency of the reaction to go forward. Spontaneous reactions have higher initial free energies and the products have lower free energies.

The following equation: $\Delta G° = \Delta H° - T\Delta S°$, in which $\Delta S°$ is the change in entropy or randomness, indicates that free energy depends on three factors: enthalpy, entropy, and temperature.

To understand how you push a reaction forward, look at the preceding equation and each of the factors. Exothermic reactions ($\Delta H° < 0$), or reactions that increase entropy (producing products in a more random phase), tend to make ΔG more negative. In addition, increasing the temperature of a reaction increases the magnitude of the $T\Delta S°$ term and makes it easier to surpass the activation energy barrier. Both of these effects help to push the reaction forward and result in a more negative ΔG.

Another relationship:

$$\Delta G° = -RT \ln K_{eq}$$

in which $K_{eq} = [\text{products}]/[\text{reactants}]$. This relationship quantifies the relationship between $\Delta G°$ and the size of the equilibrium constant. This information is important. You should be able to predict that the larger the negative value of $\Delta G°$, the larger K_{eq}. If product molecules predominate at the reaction equilibrium, then K_{eq} is high. This favorable formation of product is then reflected in a low $\Delta G°$.

D. REACTION ORDER

The order of a reaction describes how the concentration of the reactants affects the rate of the reaction. For the reaction:

$$mA + nB \rightarrow \text{product} \qquad \text{rate} = k\,[A]^m[B]^n \qquad (k = \text{rate constant})$$

The order of the reaction = m + n. Become familiar with the following types of reactions:

Zero Order: The rate of the reaction is constant and independent of the reactant concentration. An example is an enzyme-catalyzed reaction that is under conditions of substrate saturation (enough substrate so that the enzyme is working as hard as possible).

First Order: The rate depends on the concentration of one of the reactants.

Second Order: The rate depends on the concentration of two of the reactants. The reactants can be the same molecule (A + A → product) or two different molecules. Most reactions that you will encounter are first or second order.

Alkanes and their Chemistry, Stereochemistry, and Reaction Mechanisms

I. Alkanes

A. EMPIRIC FORMULA

Alkanes are carbon compounds consisting of single bonds (σ) between each carbon. They have the general formula: $\mathbf{C_nH_{2n+2}}$. Alkanes are described as saturated because the carbons of the alkane are bound to as many hydrogens as possible. A review of the naming of alkanes is provided in the Appendix.

When n > 3, it is possible for several molecules to have the same atoms but different connections. These compounds are **structural isomers** of one another. In Figure 2-1, note that the structural isomers butane and 2-methylpropane have a general formula of C_4H_{10} but different connectivity. **Structural isomers usually differ also in their physical properties** (e.g., boiling and melting points, solubilities, and densities) and spectroscopic properties [e.g., nuclear magnetic resonance (NMR) and infrared (IR) spectroscopy].

B. PHYSICAL PROPERTIES

The states of alkanes at room temperature are as follows:

n = 1–4: gases
n = 5–17: liquids
n > 18: solids

Alkanes are **highly nonpolar** and insoluble with water. They are very effective solvents for nonpolar compounds. Polar compounds cannot be dissolved in alkanes.

The melting and boiling points of these compounds are relatively low. As the carbon chain length increases, the melting and boiling points increase because of the increase in the **Van der Waals** intermolecular interactions. These interactions are caused by the transient dipoles that form as electrons move in their orbitals within a molecule. These dipoles can induce dipoles to form in other molecules, and weak dipole–dipole attractive interactions occur between these molecules. Remember that melting and boiling points are directly related to intermolecular interactions.

With increased branching, boiling point decreases because of fewer and weaker Van der Waals interactions. The branches act like arms to increase intermolecular distances, thus decreasing the strength of Van der Waals interactions.

Melting point is more complex. In many cases, melting temperature is higher because branching enhances symmetry, which in turn enhances the ability of the molecules to form crystal lattices. As an example, cycloalkanes have high melting points because of their high degree of symmetry.

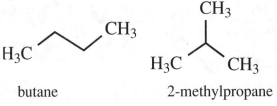

butane   2-methylpropane

Figure 2-1.   Two structural isomers.

Combustion of alkanes involves the complete oxidation of the alkane to form CO_2 and H_2O. These processes are exothermic. **More heat is released from a longer carbon chain** because there are more C, H, and O atoms in the molecule. Between isomeric forms, the one that releases more energy is the less stable isomer.

C.   CONFORMATION

Remember that the atoms about a single bond are free to rotate. **Rotational isomers are also known as conformers and are generated by rotating substituents about single bonds.** Figure 2-2 shows the rotational isomers for butane. You can generate these isomers by looking down the long axis of a molecule and seeing how the substituents are positioned with respect to one another. Realize that for a given alkane, certain rotational isomers are more energetically favorable than others.

Of the different rotational isomers possible, the basic forms—**anti, gauche,** and **eclipsed**—are shown in Figure 2-2. The anti and gauche forms are known as **staggered** conformations, because the bonds emanating from the two carbons into the plane of the paper do not overlap as they do in the eclipsed conformer.

The stability of conformers is dictated by two factors: steric interactions and eclipsing strain. Steric interactions occur when the atoms get too close and "bump" into one another. Eclipsing strain occurs when the bonds and substituents overlap.

Moving from left to right in Figure 2-2, the forms are shown from lowest to highest energy. It is important to realize that the percentage of time a molecule spends in each conformation is directly related to the energy of the conformation; that is, the highest energy conformations are the least populated. The eclipsed form (far right) is the least stable and the highest in energy because of the proximity of the methyl groups and the fact that it is an eclipsed conformation. The anti form is the most stable because of the large separation between the methyl groups in a staggered conformation. The eclipsed forms are less stable than any staggered form, given the overlap of the bonds and the substituents.

D.   CYCLOALKANES

Ring strain is caused by the geometry of the ring and the fact that sp^3 carbons normally assume an angle of 109.5°. The more this angle deviates from this value, the greater the ring strain. Ring

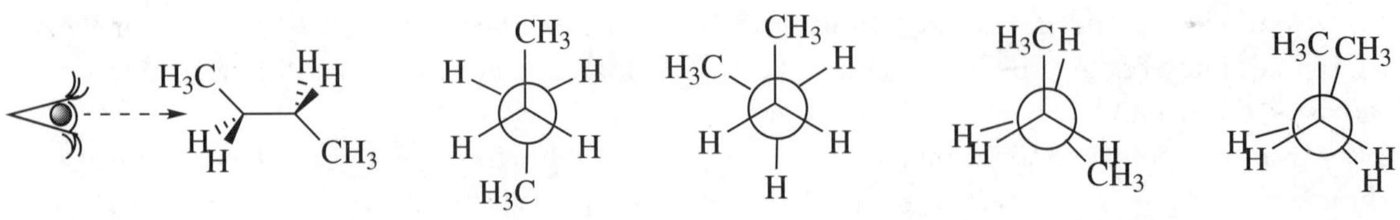

ANTI   GAUCHE   ECLIPSED

Figure 2-2.   Rotational isomers of butane.

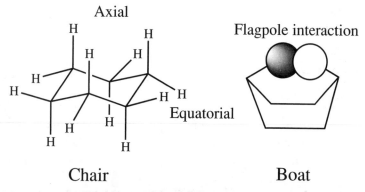

Axial

Flagpole interaction

Equatorial

Chair

Boat

Figure 2-3.   Chair and boat forms of cyclohexane. The circles in the boat form represent substituents.

strain can be measured by examining the energy released per CH_2 group on combustion. Cyclopropane has the greatest and cyclohexane has the least energy released per CH_2 group. Thus, the three-membered ring of cyclopropane is highly strained and cyclohexane has no ring strain.

Cyclohexane is unique because it is possible to have all of the carbons perfectly tetrahedral and all of the bonds are staggered. This conformation is called the **chair.** In the chair conformation in Figure 2-3, note that there are two different orientations for substituents to attach: axial (along the vertical axis) and equatorial (along the belt line of the ring). Cyclohexane can also be in a **boat** form, but this conformation is less stable because of repulsive flagpole interactions.

The concept of **ring flip** is illustrated in Figure 2-4. The form of the chair at left contains two groups that are close to one another. The repulsive interactions between groups lead to ring flips, which spread bulky groups to equatorial and more distant locations. In Figure 2-4, 1,3-diaxial interactions are the cause for the placement of larger substituents in the **equatorial** (around the equator) positions instead of the **axial** (vertical axis) positions. This occurs as a result of flips in the ring. These flips occur constantly and are in equilibrium based on their relative energies. Thus, the form with the larger substituent in the equatorial position would have the highest concentration.

II.   Stereochemistry

A.   IMPORTANCE

Many reactions occur **stereospecifically** (give different products depending on the stereochemistry of the substrate) or **stereoselectively** (yield one stereoisomer more than another). Thus, it is important to be able to name, classify, and identify stereoisomers. **An important example involves thalidomide,** the drug used in the 1960s to alleviate morning sickness. Two

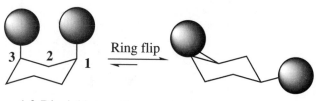

3   2   1

Ring flip

1,3-Diaxial interactions

Figure 2-4.   Ring flip.

stereoisomers were coadministered. One of the stereoisomers provided the intended relief, whereas the other was later found to cause severe birth defects.

B.   TERMS AND DEFINITIONS

Become familiar with the following terms:

Stereoisomers: isomers that have identical connectivity but different spatial configurations; two types are **enantiomers** and **diastereomers.**
Enantiomers: Stereoisomers that are nonsuperimposable mirror images of each other.
Diastereomers: Stereoisomers that are not enantiomers.
Asymmetric carbon: An sp^3 carbon with four different substituents, often referred to as a **chiral carbon.** Stereoisomers are formed by changing the orientation of the substituents of the chiral carbon.
Optically active compounds: Compounds that rotate plane-polarized light are optically active (chiral). Compounds designated "*l*" rotate the plane of the polarized light counterclockwise. Compounds designated "*d*" rotate the plane of the polarized light clockwise.

As a **general rule of thumb:** If you can draw an internal mirror plane through a molecule (drawing the plane down the center of an atom is allowed), the molecule is **not** optically active and is called **achiral.**

C.   ENANTIOMERS

Your hands are an example of enantiomers on your body: they are mirror images and you cannot superimpose them. In organic chemistry, if a compound contains a single chiral carbon, its enantiomer can be drawn by switching any two substituents on the chiral carbon. The compounds then become nonsuperimposable to one another.

1.   ABSOLUTE STEREOCHEMISTRY

R and S are used to identify and label chiral carbons. The assignment of R or S is determined by the priority of the substituents on the chiral carbon. Priorities are determined by the following process:

- Look at the atom attached to the carbon. Rank by atomic number.
- If there is a tie, go to the next atom and rank by atomic number.
- Look for a point of difference such as a branch point. If there still is a tie, go to the next atom. Keep going until there is a higher-ranking atom. An example is shown in Figure 2-5.
- Look for multiple bonds. Caution: they are counted as if they were the same number of singly bonded atoms (i.e., a $CH\text{-}CH_2$ double bond counts as four carbons and three hydrogens). See the examples in Figure 2-6.

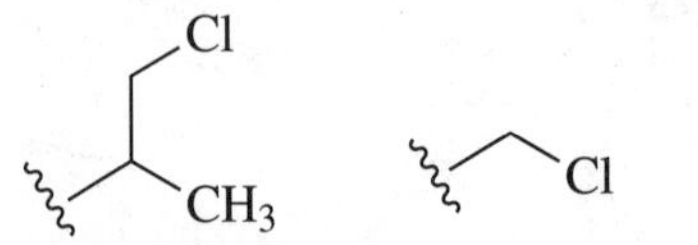

Lower priority   Higher priority

Figure 2-5.   Assigning priority for a branch point.

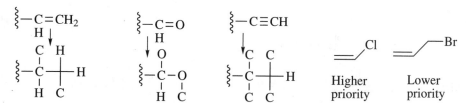

Figure 2-6. Priority assignments for multiple bonds.

- Remember to look for rankings according to atomic weight and **not** the total weight of the substituent group.

a. Steering Wheel Approach

Place the lowest-priority substituent (often hydrogen) into an imaginary steering column. Look straight at the wheel and see if the other three ranked substituents go from high to low priority clockwise (R) or counterclockwise (S). Recognize that if a chiral center is assigned R, its mirror image is S.

Figure 2-7 shows a chiral molecule and its enantiomers. The geometry of the enantiomers (at left) reveals that they are nonsuperimposable mirror images of each other. Using the steering wheel approach to assign priority of substituents, compound **A** shows an R conformation and compound **B** shows an S conformation.

b. Fischer Projections

Fischer projections can also be helpful in assigning absolute stereochemistry. Think of the substituents along the horizontal bonds as a bow tie. To assign R or S, rank the substituents as described previously. Put the lowest-priority group at the top by rotating the whole projection by 180° (rotations by 90° or 270° result in the representation of the other enantiomer), or hold one atom and rotate the other three atoms.

In Figure 2-8, the Br is held and the other three positions are rotated to place the lowest-priority H at the top. With this action, you are just spinning the chiral carbon about one of the bonds. Assign the compound by examining the orientation of the groups from high to low priority (R—clockwise, S—counterclockwise).

2. PHYSICAL PROPERTIES

Enantiomers are **optically active.** One enantiomer rotates plane-polarized light in one direction and the other rotates it in an equal but opposite direction. **All other physical (e.g., melting and boiling points and solubilities) and spectroscopic (e.g., NMR, IR) properties of enantiomers are identical.** Thus, enantiomers are hard to separate, purify, and characterize.

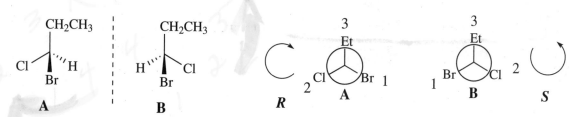

Figure 2-7. Assigning R and S absolute stereochemistry to two enantiomers, A and B, using the steering wheel approach.

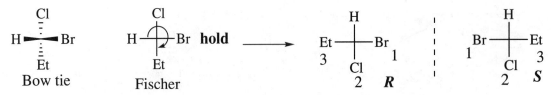

Figure 2-8. Using Fischer projections to assign absolute stereochemistry.

3. RACEMIC MIXTURE

A racemic mixture is a solution with an equal concentration of two enantiomers. The solution does not rotate plane-polarized light; it is rotated in one direction by one enantiomer and rotated to an equal degree in the opposite direction by the other enantiomer.

D. DIASTEREOMERS

1. STEREOCHEMISTRY

Diastereomers, defined as stereoisomers that are not enantiomers, are compounds with two or more asymmetric centers.

Remember: For n-chiral centers in a molecule, there are 2^n possible stereoisomers.

Study Figure 2-9. Note that some compounds are nonsuperimposable mirror images and others are not. Be sure you understand which pairs are enantiomers and which are diastereomers. (Answers are provided in the legend.)

2. PHYSICAL PROPERTIES

Diastereomers have **different** melting and boiling points, solubilities, and spectroscopic properties. Thus, diastereomers can be isolated, purified, and characterized. Diastereomers also have different, but not opposite, optical activities.

E. MESO COMPOUNDS

A meso compound has asymmetric centers and an internal mirror plane of symmetry. Meso compounds are not optically active and do not possess enantiomers. Try to draw a line of symmetry through a compound to see if it is a meso compound. Compounds A and B in Figure 2-9 are both meso compounds.

Figure 2-10 illustrates two sets of compounds with R and S configurations. Compound **A** is a meso compound because you can draw a mirror plane down the middle of the molecule. **B** is a mirror image of **A** and is also a meso compound. Note that both **A** and **B** are superimposable. Compounds **A** and **B** are identical because they just flip onto one another. Compounds **C** and **D** are not meso compounds; they are enantiomers. **(C,A)** and **(D,A)** are pairs of diastereomers.

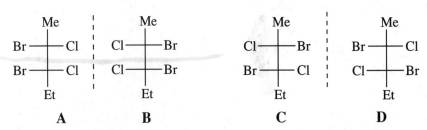

Figure 2-9. Enantiomers and diastereomers. Compounds **A** through **D** each have 2 chiral centers and 2^2, or 4, stereoisomers. **(A,B)** and **(C,D)** are pairs of enantiomers. **(A,C)**, **(A,D)**, **(B,C)**, and **(B,D)** are pairs of diastereomers.

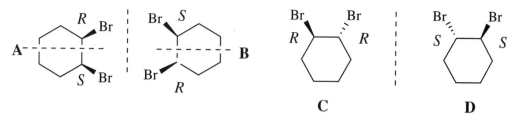

Figure 2-10. Examples of meso compounds, enantiomers, and diastereomers.

III

Free Radical Halogenation of Alkanes

A. THE BASICS

Free radical halogenation of alkanes involves initiation (formation of halogen radical), propagation (chain reaction resulting in product and halogen radical), and termination (two radicals coming together to form a bond) steps. Once the initial radical is formed, the rest goes on by itself. Lower-energy radicals are more selective (Br• is the most selective radical), which means that the lower-energy halogen radicals preferentially remove hydrogens that form lower-energy (more stable) alkyl radicals.

B. TYPES OF BOND CLEAVAGE

Heterolytic cleavage involves the breaking of a bond such that both electrons from the bond are taken by one of the atoms.

Example 2-1: $H_2O \rightarrow H^+ + OH^-$

Homolytic cleavage involves the breaking of a bond such that one electron goes to one side and the other goes to the other side. Unlike heterolytic cleavage, the products are not charged. The radicals that are formed are very reactive because they are electron deficient and want a full outer shell.

Example 2-2: HO-OH (hydrogen peroxide H_2O_2) $\rightarrow$ HO• + •OH

C. REACTION SEQUENCE

The steps of free radical reactions are detailed in Figure 2-11 and as follows.

Initiation: The halogen radicals are formed by reacting with light or heat. This is highly endothermic and ends at point A on the energy graph in Figure 2-11.

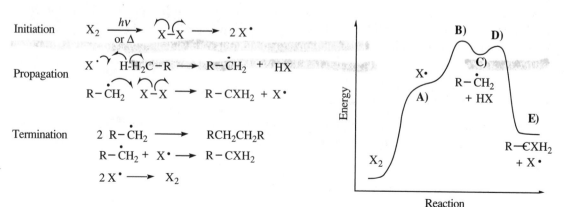

Figure 2-11. Free radical halogenation: the reactions and energy diagram. *Single-headed arrows* identify single electron movements.

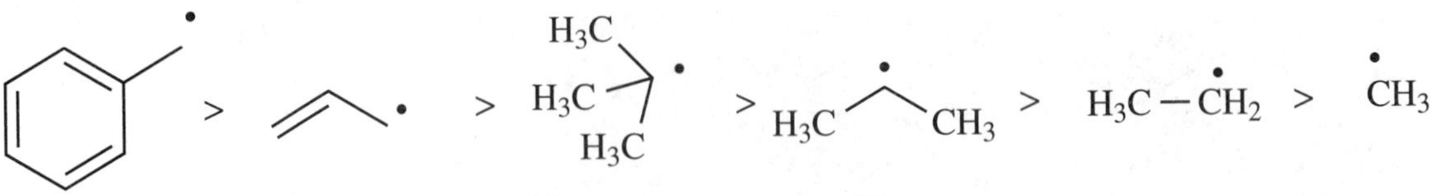

Figure 2-12. Stability of radicals.

Propagation: The halogen radical then pulls off a proton from the alkane (homolytic cleavage) to form HX and the alkyl radical. The alkyl radical then splits up another X_2 molecule and the propagation steps go on until one of the starting materials runs out or the radicals are consumed by termination steps.

    Note that points **A** to **C** on the graph involve formation of the alkyl radical. This is highly endothermic and requires energy input to overcome the activation energy barrier. After **C** is formed, the formation of the alkyl halide and the halogen radical (**C to E**) is highly exothermic, with a relatively small activation energy barrier from points **C** to **D**.

Termination: Combination of any two radicals creates a nonradical species and propagation steps stop. This is best depicted at point **E**. Termination steps usually are energetically favorable.

D. ENERGY AND STABILITY OF RADICALS

Many questions can be answered by understanding stability concepts. Remember from the review of acids and bases in Chapter 1 that spreading the charge increases stability. Radicals are electron deficient, and so the attachment of electron-donating substituents, like carbons, leads to greater stabilization **(inductive effect, spreading the charge).** In Figure 2-12, the stability of radicals is ranked high to low. Note that **radicals are stabilized by delocalization and resonance.**

    Radicals can be stabilized by **delocalization.** For alkyl radicals, the p orbital of the radical can line up with an σ C-H bond to give the resonance forms shown in Figure 2-13. If R is an alkyl group, several more of these types of resonance forms can be drawn. **Thus, the more substituted alkyl radicals are more stable.** Furthermore, the alkyl groups can be considered electron-donating substituents that can stabilize the electron-deficient radical. **In all cases, stability is enhanced by spreading the electrons.**

    For the halides (high to low reactivity), F• > Cl• > Br•. I• is unreactive. This order relates to the fact that F is highly electronegative and is very willing to form a bond. F• is high enough in energy to pull off any proton to form any alkyl radical (primary to benzyl). Br•, on the other hand, is lower in energy and is thus **more selective** of which H atom it pulls off. Given a choice, the Br• species pulls off the H atom that leads to the most stable alkyl radical because it is the easiest H atom to remove. Halogenation and analogous radical reactions are successful only with Cl• or Br•. Unfortunately, F• is too reactive and I• not reactive enough for these reactions.

E. STEREOCHEMICAL ASPECTS

If the alkane is a pure enantiomer, the resulting product will be a mixture of two enantiomers. Halogen attack can occur from either the top or bottom face to give two different products. Thus,

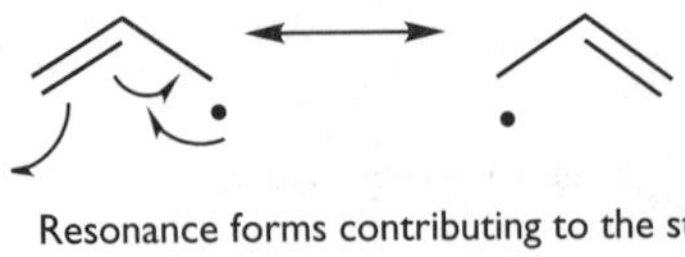

Figure 2-13. Resonance forms contributing to the stability of radicals.

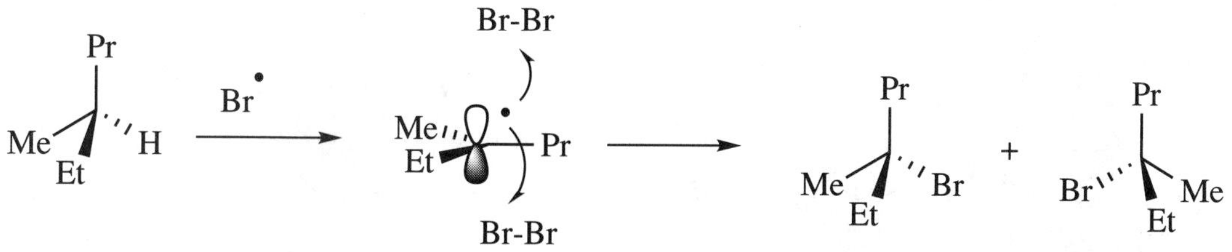

Figure 2-14. Enantiomers produced from free radical reaction. Do not memorize this process; just understand that it occurs.

racemization occurs. Figure 2-14 illustrates the production of two enantiomers from free radical reaction.

IV

Nucleophilic Substitution and Elimination

To understand these critically important mechanisms fully, you must know why they occur, how they occur, and what products they produce.

A.   GENERAL IDEAS

Reactions often occur (bonds break and new ones form) between a molecule that is electron rich (δ^-) and one that is electron deficient (δ^+). Such is the case in radical halogenation, because radicals are the electron-deficient species, as well as with nucleophilic substitutions and eliminations.

Substitutions and eliminations involve the reaction between an electrophile (electron-deficient species) and a nucleophile (electron-rich species). **Eliminations result in alkenes and substitutions result in alkanes with different substituents.** Note also that electron-rich and electron-deficient species exist, because of the formation of dipoles caused by differences in electronegativity. The types of substitutions and eliminations can be distinguished by mechanism and product distribution.

The substitutions are commonly known as S_N1 and S_N2 reactions. The eliminations are commonly known as E1 and E2. **Substitution and elimination reactions compete in many instances.** The conditions of the reaction and compounds involved favor one type of reaction over another.

B.   S$_N$1

An S_N1 reaction involves two steps and is a **unimolecular reaction** (first-order reaction; see Chapter 1). The **first step** involves the **dissociation of the leaving group (electron-accepting species) from the substrate, resulting in a carbocation.** The carbocation is considered an intermediate. **Intermediates** are characterized as high-energy valleys in the energy diagram that either go forward to create product or degrade back to starting materials. The **second step** involves the **attack of the carbocation by a nucleophile from either face of the sp^2 hybridized carbocation.**

In Figure 2-15, the first step of the S_N1 reaction involves the formation of the carbocation intermediate. The second step involves the nucleophilic attack of the carbocation, leading to a substituted product. Note that the nucleophile can attack from either face. The energy graph (Figure 2-15, at right) shows the formation of the carbocation intermediate (C+).

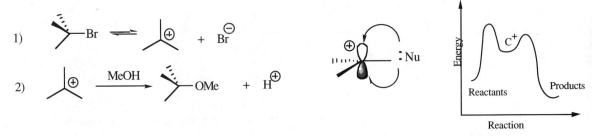

Figure 2-15. S_N1 reaction mechanism. Note the two steps, the ability of the nucleophile to attack from either face, and the energy diagram depicting the reaction.

1. RATE-DETERMINING STEP

The step with the highest activation energy determines the rate of the reaction. In this process, the formation of the carbocation is the rate-determining step. Because the formation of a more stable carbocation requires less energy, the stability of the carbocation directly affects the rate of the reaction.

2. CARBOCATION STABILITY

Carbocation stability (Figure 2-16) parallels radical stability (see preceding section III) and for similar reasons.

Increased substitution results in more groups that can donate electrons (electron induction) to stabilize the carbocation (remember: spread out the charge, increase stability). The benzyl and the allyl are especially stable because of **delocalization** of the positive charge through resonance.

3. CARBOCATION REARRANGEMENT

Figure 2-17 shows that carbocations can migrate through shifts to form a more stable carbocation.

4. FIRST-ORDER RATE

Rate of the reaction depends only on the concentration of the substrate. The strength and concentration of the nucleophile are not as important, because nucleophilic attack is **not** the rate-limiting step.

5. STEREOCHEMICAL CONSIDERATIONS

If you start with an optically pure material (only one enantiomer present), the result will be a mixture of stereoisomers. This change occurs because the carbocation is sp^2 hybridized and can be attacked by the nucleophile from either face, resulting in two different stereoisomers.

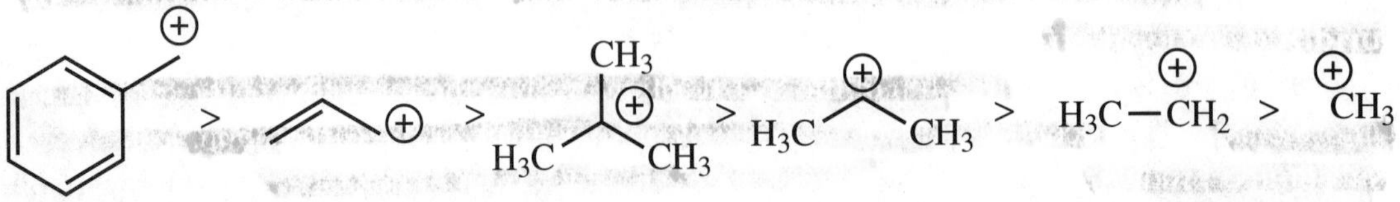

Figure 2-16. Carbocation stability.

SECTION I · ALKANES AND THEIR CHEMISTRY, STEREOCHEMISTRY, AND REACTION MECHANISMS

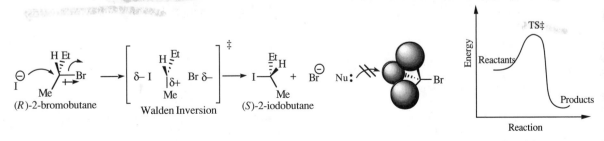

Figure 2-17. Carbocation rearrangement to form the most stable carbocation.

C. S$_N$2

S$_N$2 is a **one-step, bimolecular** (second order; see Chapter 1) reaction. Refer to Figure 2-18 and follow the steps of the reaction closely.

The nucleophile approaches the electrophile along the line of the electrophile leaving group bond, the dipole formed by the electronegative leaving group having made the carbon of the bond δ^+. The nucleophile attacks as the leaving group leaves. This action correlates with the transition state (TS) shown on the energy graph in Figure 2-18. During this stage, the electrophile is flattened out like an umbrella in the wind. The new bond between the nucleophile and the electrophile is formed and the electrophile leaving group bond is broken. Because the reaction is one-step, there is **no rate-determining step.**

1. STERIC CONSIDERATIONS

If the electrophile has too much steric bulk (large alkanes or branched alkanes), the reaction occurs slowly or not at all. The same holds true for bulky nucleophiles.

2. STEREOCHEMICAL CONSIDERATIONS

An inversion of stereochemistry follows the nucleophilic attack. Optical activity is maintained, although the product will have a different optical rotational value.

3. SECOND-ORDER RATE

Unlike S$_N$1, the **rate depends on the concentration of both the electrophile and the nucleophile.**

D. S$_N$I AND S$_N$2: DIFFERENCES AND PREFERENCES

Consider the following to predict which substitution mechanisms are favored.

1. ELECTROPHILE

Both reactions prefer strong electrophiles. In terms of substitution, however, opposite trends are at work. **S$_N$2 prefers less substituted electrophiles (primary or secondary carbons)**

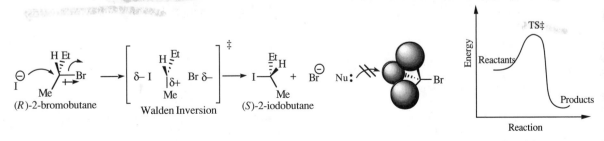

Figure 2-18. S$_N$2 reaction mechanism. Note the single step, transition state, geometry of the nucleophilic attack, and the energy diagram depicting the reaction.

so the attacking nucleophile can approach the electrophile. Tertiary electrophiles do not undergo S_N2.

S_N1 **prefers more substituted electrophiles (tertiary > secondary > primary)** because of the stability of the carbocations that are formed after dissociation of the leaving group.

2. NUCLEOPHILE

Both reactions work well with good nucleophiles, but **the S_N2 reaction is more dependent on the strength of the nucleophile.** Remember that the rate-determining step of the S_N1 reaction is the formation of the carbocation, which is independent of the nucleophile.

What makes a good nucleophile?

a. **Good nucleophiles are usually good bases.** Do not think that all strong bases are good nucleophiles. The only possible generalization is that for a group of nucleophiles in which the attacking atom is the same: the stronger base is the better nucleophile.

b. **Charged nucleophiles are always more powerful than their uncharged counterparts.**

c. **The halogens are good nucleophiles because of their ability to polarize (fluffiness).** The electron cloud around the nucleus is large and the outer electrons are shielded from the positive charge of the nucleus. This makes it easy for larger atoms to donate electrons to electrophiles.

Good examples of nucleophiles are as follows:

- Conjugate bases of weak acids: ^-CN, RO^-, HO^-, RS^-, among others
- Halogens: I^-, Br^-, Cl^-
- Neutral molecules with lone pairs: RNH_2, ROH, H_2O, RSH
- Carbon nucleophiles: R_2CuLi

3. LEAVING GROUP

Both S_N2 and S_N1 reactions prefer good leaving groups. **For the S_N1 reaction, a better leaving group lowers the activation energy of the rate-determining step.**

What makes a good leaving group?

Good leaving groups are conjugate bases of strong acids (i.e., weak bases). A good leaving group can stabilize the extra electron it gains from dissociation from the carbon. Thus, groups such as I^-, Br^-, are good leaving groups.

^-OH is considered a poor leaving group (strong base), but H_2O is considered a good leaving group (weak base). Figure 2-19 demonstrates how you can improve a leaving group by modifying it. In this example, the poor leaving group, the hydroxyl, is protonated to make it a weaker base and a better leaving group. This reaction then proceeds by way of an S_N1 mechanism. Thus, protonation can allow for displacement of OH by an **S_N1-type reaction.** S_N2 does not work by this method.

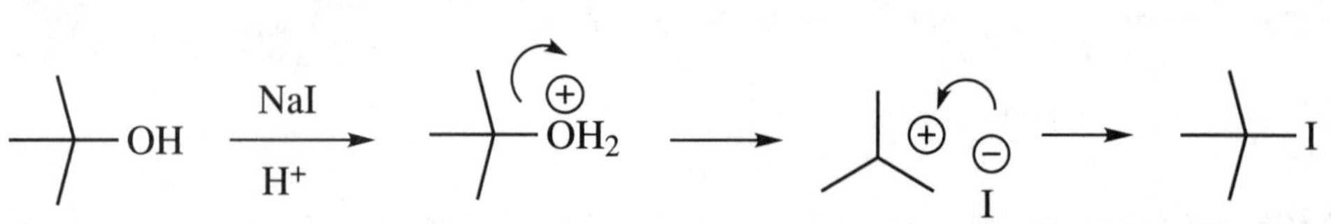

Figure 2-19. Improving a leaving group by protonation, allowing an S_N1 reaction.

4. STEREOCHEMICAL IMPLICATIONS

If the starting material of an S_N2 reaction is chiral, the product will be chiral **with inversion of stereochemistry.** For S_N1, the product will be a racemic mixture because of the formation of the carbocation and nucleophilic attack occurring in two different directions.

5. SOLVENT EFFECTS

Both reactions favor polar solvents. For **S_N2, aprotic solvents** (no protons that can be involved in hydrogen bonding) are favored. For **S_N1, protic solvents** (good donors of protons for H-bonds), such as alcohols, are favored because of their ability to stabilize the anion and cation products of the rate-limiting step of the reaction.

6. MECHANISMS

A summary of mechanisms for the substitution reactions follows.

- S_N1 goes through a mechanism that involves an **intermediate,** whereas S_N2 goes through a single step with a transition state.
- The **rate** of the S_N1 reaction **depends on substrate concentration** but not on nucleophile concentration or strength.
- The **rate** of the S_N2 reaction **depends on both substrate and nucleophile** concentration and strength.
- The **products** of S_N1 reactions can be the result of **carbocation rearrangements.** These rearrangements do not occur for S_N2.

E. E1

Like S_N1, E1 is also a **two-step, unimolecular** reaction. Refer to the example in Figure 2-20.

The first step involves the dissociation of the leaving group from the substrate, resulting in a carbocation intermediate (just like S_N1). The second step involves the removal of a proton by a base and the formation of a double bond. The energy diagram for this process is analogous to the S_N1 energy diagram. **The rate-determining step is again the carbocation formation.**

1. REARRANGEMENTS

As in the S_N1 reaction, the carbocations can rearrange to a more stable carbocation.

2. ALKENE STABILITY

On elimination, several different double-bonded molecules can result. The more stable double bond forms preferentially. In most cases, **the product with the increased number of**

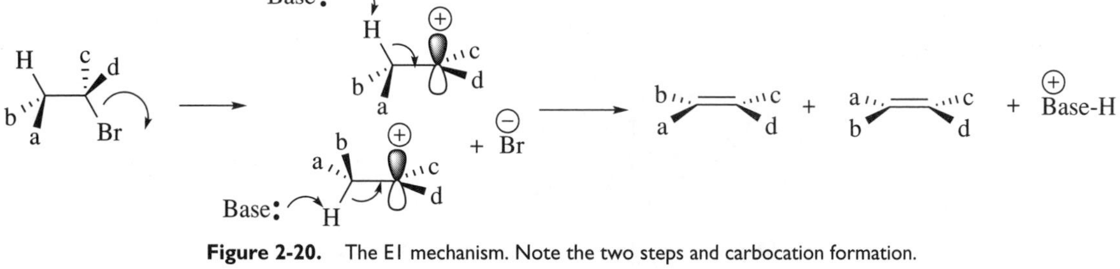

Figure 2-20. The E1 mechanism. Note the two steps and carbocation formation.

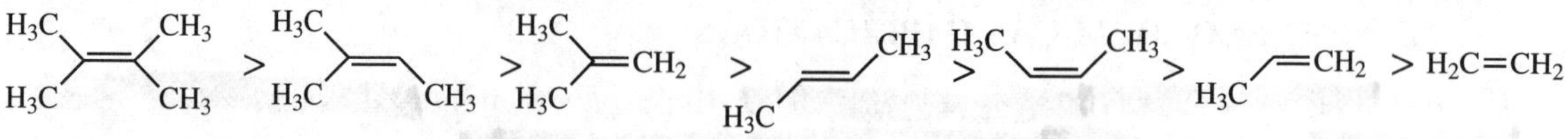

Figure 2-21. Alkene stability in EI reactions.

substituents on the double bond (higher substitution) is favored. This arrangement increases stability. In addition, the ***trans* isomer is more stable than the *cis* form because of eclipsing strain.** (See the Appendix for a review of *cis/trans* designations.)

Figure 2-21 shows which alkene products are most stable in E1 reactions (listed from high to low stability).

3. RATE

Like S_N1, the rate of E1 depends on the concentration of the substrate and is independent of the concentration or strength of the base. **In general, E1 is outcompeted by S_N1 and is not used as an elimination reaction.**

F. E2

Like S_N2, E2 is a **single-step, bimolecular** reaction. Refer to the example in Figure 2-22.

The base deprotonates the substrate when the proton is **antiperiplanar** to the leaving group. In this orientation, the proton is on the positive end of the dipole of the molecule. Electrons move from the broken C-H bond to the C-C bond to form a double bond as the leaving group departs. The energy diagram resembles the diagram for the S_N2 reaction.

1. RATE

The rate of the E2 reaction depends on the concentration of the substrate and base along with the strength of the base (stronger bases are more effective). By increasing the steric bulk of the base, fewer S_N2 side reactions occur.

2. STEREOCHEMICAL CONSIDERATIONS

Given the nature of the antiperiplanar orientation of the substrate, specific products can be expected. E1 gives a mixture of products, but in E2, **rearrangement does not occur.** Remember that E2 is a one-step reaction.

Key point: E2 is a more effective elimination reaction than E1.

G. TAKE-HOME POINTS TO UNDERSTAND

1. SUMMARY CHART

The chart in Figure 2-23 sums up the reactions discussed in this chapter and the conditions favoring the reaction types. Note that you can control the reaction path by controlling the conditions of the reaction.

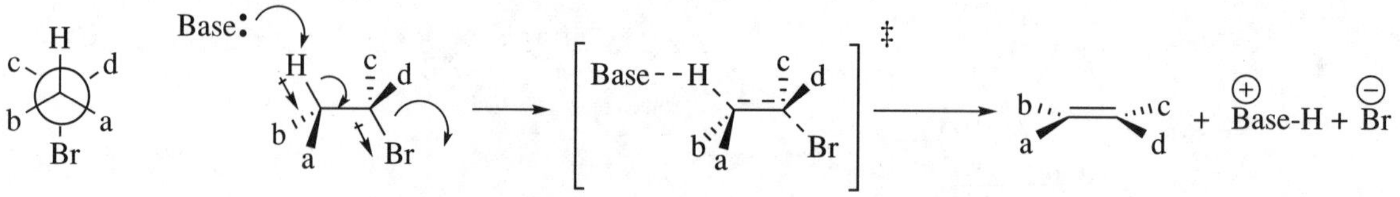

Figure 2-22. E2 mechanism. Note the transition state and double bond formation.

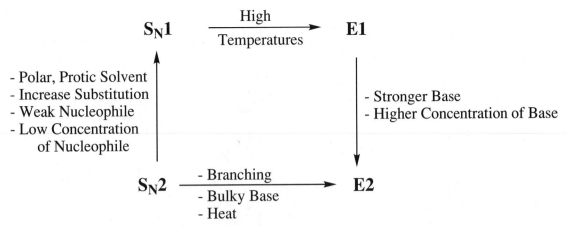

Figure 2-23. Summary chart of reaction types.

2. MORE GENERAL RULES

a. Primary alkyl halides undergo S_N2 or E2, depending on the base.
b. Secondary alkyl halides undergo S_N2 with weak bases, E2 with strong bases, and S_N1 when protic solvents are used.
c. Tertiary alkyl halides undergo S_N1 with no S_N2. E1 also occurs, but is more of an annoying side reaction that ruins yield. E2 is the only observed reaction with a strong base.

Alkenes and Alkynes ③

I. General Information

A. EMPIRIC FORMULA

Alkenes are hydrocarbons that possess a double bond between adjacent carbon atoms in the hydrocarbon backbone. The double bond acts as the dominant functional group of the molecule. **For alkenes, the general formula is: C_nH_{2n}.** (See the review of nomenclature for alkenes and alkynes in the Appendix.)

Alkynes are hydrocarbons that possess a triple bond between adjacent carbon atoms in the hydrocarbon backbone. The triple bond is the dominant functional group. **For alkynes, the general formula is: C_nH_{2n-2}.**

B. STRUCTURE AND PHYSICAL PROPERTIES

Like alkanes, the **melting and boiling points of alkenes and alkynes increase with chain length.** Unlike alkanes, **no free rotation** occurs about the double bond (alkene) or triple bond (alkyne) because of the presence of the π-bonds. The carbon-carbon double bond is shorter and stronger than a carbon-carbon single bond owing to the π-interaction and increased σ-character of the σ-bond. Similarly, a carbon-carbon triple bond is stronger and shorter than a carbon-carbon double bond.

The stability of the multiple bond increases with substitution. For double bonds, *cis* **is less stable than *trans*** because of eclipsing interactions of the substituents. The carbons of multiple bonds are slightly more electronegative than sp^3 carbons. The triple-bonded carbon is slightly more electronegative than the double-bonded carbon because of the increased σ-character of the sp orbitals.

C. TYPES OF REACTIONS

Most reactions of alkenes and alkynes take advantage of the accessible electrons in the π-bond. The three basic types are as follows:

1. **Electrophilic attack.** The π-bond provides a source of electrons for electrophiles to attack. These reactions go through a carbocation intermediate or an intermediate with a partial positively charged carbon.
2. **Reductions** can occur that convert the multiple bond to a more saturated bond.
3. **Oxidations** (or additions of oxygen) can occur.

A. ELECTROPHILIC ATTACK

Multiple bonds are characterized as electron-rich species with relatively accessible electrons in the π-bond(s). Because of the nature of the π-bond, with its sideways overlap, the electron can be donated to form new σ-bonds.

1. ADDITION OF HYDROGEN HALIDES (HX ATTACK)

This reaction, in which an alkene is converted by HCl, HBr, or HI into the corresponding alkyl halide, is a good example of an electrophilic attack (Figure 3-1). The π-electrons from the double bond attack the polarized HX molecule and form a bond with the hydrogen, creating a carbocation at the more substituted carbon. The anion can then attack the carbocation from either the top or bottom face, leading to the nucleophile being attached to the more substituted position. Figure 3-1 demonstrates that even with only one starting material, two products are formed because of the nonselective attack of the carbocation by the nucleophile.

It is important to know **Markovnikov's rule: in the addition of an acid to the carbon-carbon double bond of an alkene, the hydrogen of the acid attaches itself to the carbon that already holds the greater number of hydrogens.** With some rewording, Markovnikov's rule states that **electrophilic addition to a carbon-carbon double bond involves the intermediate formation of the more stable carbocation.**

Markovnikov's rule occurs because the breaking of a double bond occurs more easily in the direction that forms a more stable carbocation intermediate. Thus, the initial electrophilic attack forms the more substituted (stable) carbocation.

2. ADDITION OF HALOGENS

This reaction is another example of electrophilic attack. Figure 3-2 shows the mechanism for addition of halogens across a double bond.

These reactions begin when electrons from the double bond attack the positive end of the halogen dipole. A three-membered ring forms, as shown in Figure 3-2. Both carbons are highly δ^+ (partially positive) because of the electron-deficient, positively charged halogen of the ring. The nucleophile attacks this intermediate from the opposite side of the ring (owing to steric hindrance). Furthermore, the attack occurs at the more substituted carbon, because it can better stabilize the partial positive charge. Thus, **Markovnikov's rule is observed.**

A similar mechanism operates for alkynes (Figure 3-3). The three-membered ring in this case is highly strained and is thus **higher in energy** (i.e., has a high activation energy barrier). In most cases, high-energy activation barriers are surpassed by heating.

B. ANTI-MARKOVNIKOV REACTIONS

These important reactions result in the nucleophile attaching to the lesser substituted carbon.

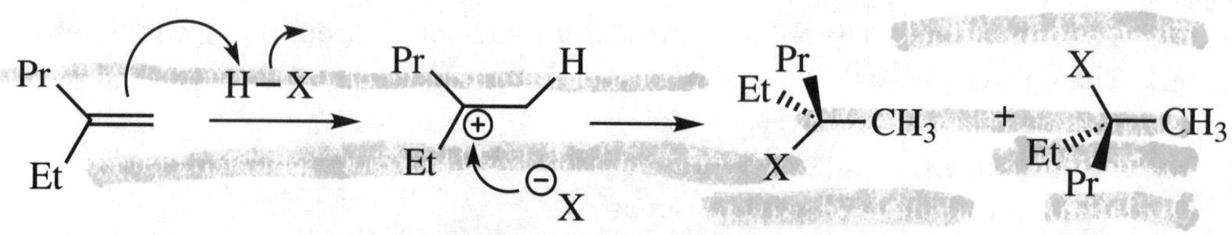

Figure 3-1. Electrophilic attack with a carbocation intermediate.

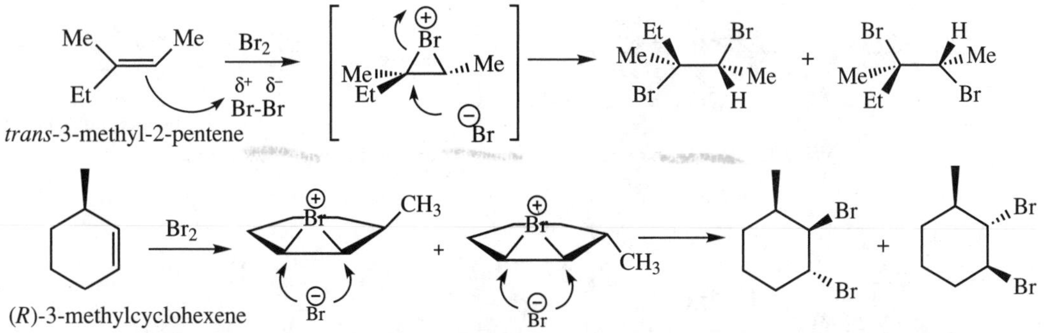

Figure 3-2. Two examples of the addition of halogen reaction. Mechanisms look complicated on first glance, but are shown for understanding, not memorization.

1. RADICAL HALOGENATION

This reaction is important because an anti-Markovnikov product is formed (Figure 3-4). Using a **peroxide** to initiate the reaction, the electron-deficient bromine radical attacks the double bond, leaving the more stable and substituted radical. The more stable radical is formed by the addition of the bromine radical to the lesser substituted site. The reaction propagates as another bromine radical is formed and continues until the reaction terminates or the starting material is consumed. This results in **an anti-Markovnikov addition of bromine to the alkene.** Even so, note that the trend of going through the more stable (lower-energy) intermediate is preserved.

2. HYDROBORATION

In this reaction (Figure 3-5), the electron-deficient boron attacks the double bond at the less substituted site, leaving the partial positive charge on the other, more substituted carbon. A hydro-

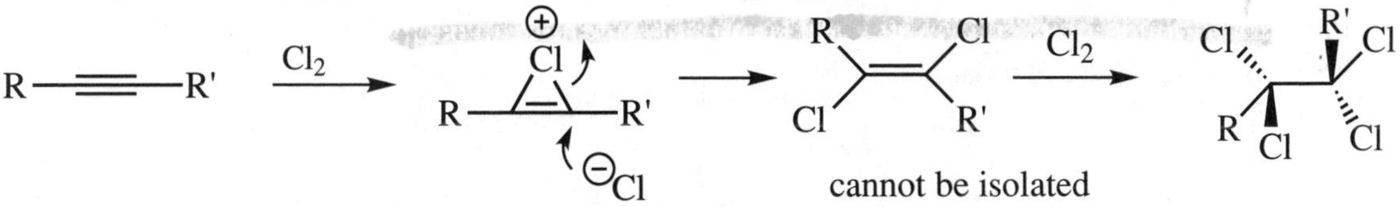

Figure 3-3. Addition of halogens across a triple bond. Mechanism shown for understanding, not memorization.

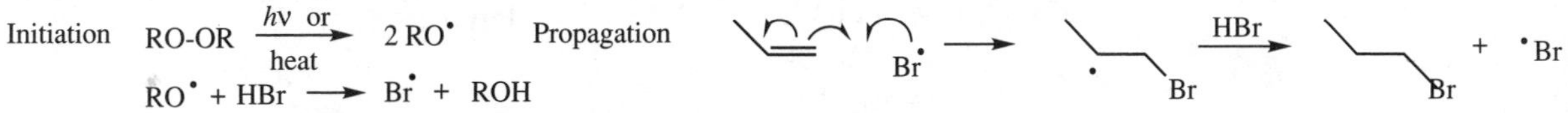

Figure 3-4. Radical halogenation, an example of an anti-Markovnikov reaction.

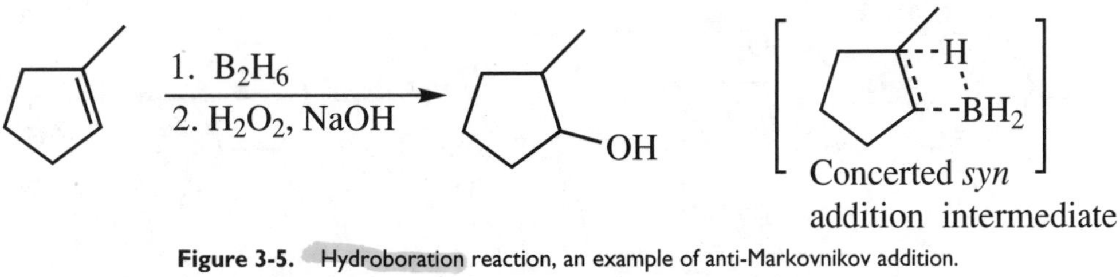

Figure 3-5. Hydroboration reaction, an example of anti-Markovnikov addition.

35

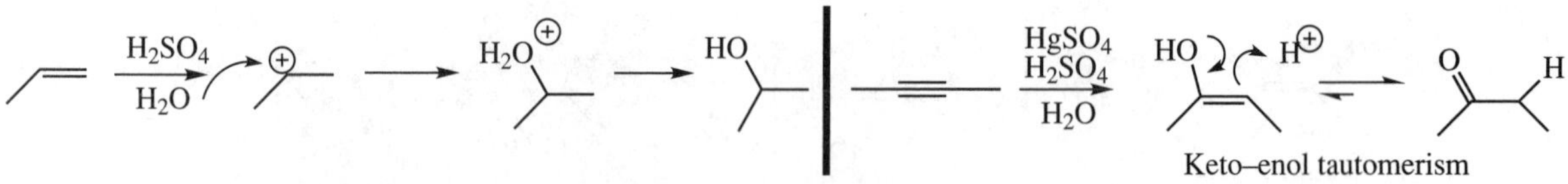

Figure 3-6. Hydration reactions for double bonds (left) and triple bonds (right).

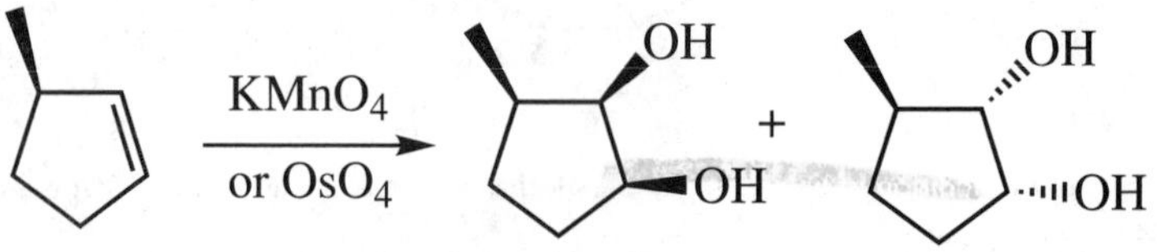

Keto–enol tautomerism

Figure 3-7. Oxidation by $KMnO_4$ or OsO_4.

gen is then delivered to this site and oxidation using hydrogen peroxide removes the boron and replaces it with a hydroxy. Thus, hydroboration represents an anti-Markovnikov addition of OH. Again, note the trend of going through the more stable (lower-energy) intermediate.

C. HYDRATION OF MULTIPLE BONDS

Water adds to the more reactive alkenes in the presence of acids to yield alcohols. This addition follows Markovnikov's rule. A hydrogen atom from water adds to the carbon of the double bond that contains more hydrogens. The -OH group from water adds to the carbon of the double bond containing fewer hydrogens. This reaction hydrates both double and triple bonds in the presence of dilute acid. In both cases, the reaction starts with the electrophilic attack of the proton, resulting in the most substituted carbocation.

Figure 3-6 shows the hydration reaction for both double and triple bonds. For the double bond, the carbocation is attacked by water and Markovnikov addition is observed. For alkynes, the carbocation is attacked by water and keto-enol tautomerism occurs. (See Chapter 6 for a review of keto-enol tautomerism.)

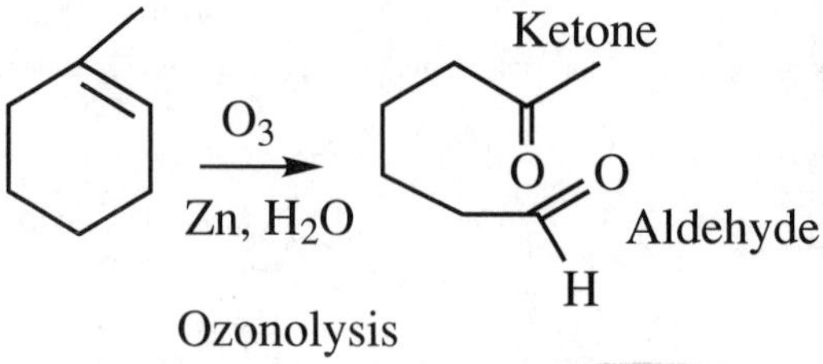

Figure 3-8. Ozonolysis reaction.

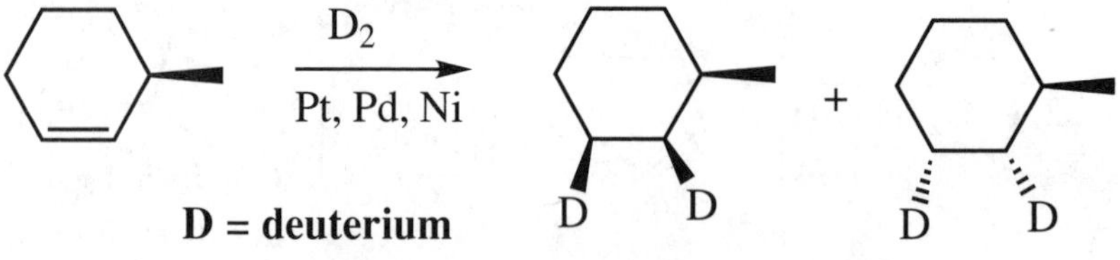

D = deuterium

Figure 3-9. Reduction of a double bond. Use of deuterium.

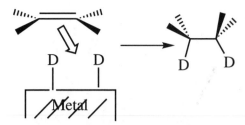

Figure 3-10. How a catalyst aids in the reduction of the double bond.

D. OXIDATIONS

To identify oxidizing agents, look for multiple oxygens in the agents themselves; $KMnO_4$ and OsO_4 are good examples. The multiple bonds are oxidized by the addition of oxygen to the bonds.

Treatment with **$KMnO_4$ or OsO_4** creates *cis* (same side) diols (Figure 3-7). If the starting material is optically active, two products form (diastereomers) because of the two different *cis* additions that can occur.

Ozonolysis (Figure 3-8) splits up the double bond and puts carbonyls in its place. This reaction is easy to recognize because of the products. Be aware that the starting materials can be linear or cyclic. Also, depending on the substitution of the double bond, an aldehyde or a ketone will result.

E. REDUCTIONS

In organic chemistry, the word **reduction** generally refers to the addition of hydrogen. In the case of multiple bonds, it means that the bonds become more saturated.

Catalytic hydrogenation of multiple bonds results in a *cis* addition of hydrogen. As shown in Figure 3-9, the location of hydrogen addition can be deduced experimentally by the addition of deuterium instead of hydrogen. It is then possible to follow the reduction process and see where the hydrogen adds.

Figure 3-10 shows the mechanism by which a metal catalyst helps reduction reactions. Presumably, the hydrogen binds to the metal catalyst and the reaction is controlled such that the substrate gets attacked on only one face.

Catalytic hydrogenation can be used to reduce alkynes as well as alkenes. The difficult part is stopping the reduction of the triple bond before complete reduction occurs. Special care is needed when reducing a triple bond to a double bond to ensure that over-reduction does not occur. Otherwise, triple bonds are reduced to the carbon-carbon single-bonded alkane.

F. FORMATION OF ALKYNES

This concept is straightforward. Recall that alkenes are formed by the elimination of hydrogen and a leaving group. To form a triple bond (alkyne), the same step is carried out twice through an E2 mechanism (Figure 3-11).

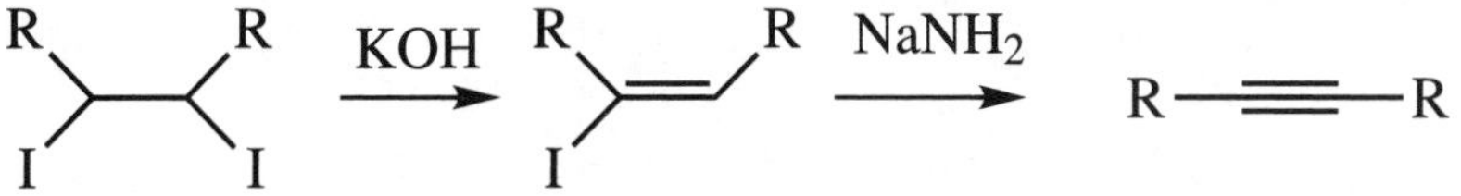

Figure 3-11. Formation of alkynes.

Benzene  ④

I. Structure and Physical Properties

A. STABILITY THROUGH RESONANCE

Conjugation refers to molecules with p orbitals next to π-bonds or an extended series of overlapping p orbitals. These conjugated structures are more stable than their unconjugated counterparts because **resonance** allows for electron **delocalization.** Thus, these compounds release less energy on combustion compared with their unconjugated counterparts.

Recall that the allylic and benzylic radicals and carbocations are highly stabilized owing to resonance. This factor determines the outcome of the reactions in which these compounds are intermediates.

Structurally, for these compounds to form conjugated structures, they must be able to line up all of their p orbitals. Thus, conformationally, these compounds favor a planar geometry. 1,3-Butadiene is a good example of a compound with planar geometry (Figure 4-1).

Once the four p orbitals for 1,3-butadiene line up, they can form molecular orbitals over the four atoms in the chain. Each p orbital contains one electron, and these four electrons fill the lower-energy–bonding molecular orbitals to capacity. This situation is energetically favorable, and is like the filling of an outer electron shell for an atom. Thus, the molecule is highly stabilized.

Benzene presents a similar molecular orbital picture (Figure 4-2). Benzene is a planar, six-membered ring with six p orbitals. These p orbitals come together to form molecular orbitals over all six atoms. The p orbital electrons fill the lower-energy–bonding molecular orbitals to capacity, creating an energetically favorable situation that makes benzene highly stabilized. Think of benzene as a flat carbon ring with a π-electron cloud above and below the ring (see Figure 4-2).

B. AROMATICITY

Aromatic compounds usually have double bonds, are stable, and usually are found as five-, six-, or seven-membered rings. In theory, aromatic compounds are thought to have cyclic clouds of delocalized π-electrons above and below the plane of the molecule.
How can you be sure that a compound is aromatic?

A compound is considered aromatic if it has (4n + 2) π electrons (n = 0, 1, 2, etc.) and is in a cyclic form. In addition, each member of that cyclic compound must be associated with at least one sp² hybridized atom. Thus, you can determine whether a particular compound is aromatic by calculating the number of its π-electrons and seeing if that value is predicted by the formula: 2, 6, 10, etc.

Benzene is an aromatic compound and contains six π-electrons. Many compounds, however, that do not look exactly like benzene are aromatic. Study the compounds in Figure 4-3 to determine whether each compound is aromatic. Use the formula provided. Each double bond represents two π-electrons. Unpaired electrons are also counted directly. Sum all of the

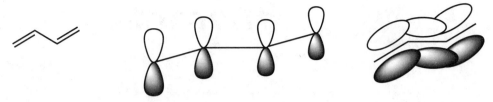

Figure 4-1. Arrangements of 1,3-butadiene. Note the structure (left), the p orbitals for the molecule (center), and a representation of the molecular orbitals (right).

Figure 4-2. Benzene: resonance structures (left), resonance hybrid (center), and π-electron cloud above and below the ring (right).

π-electrons associated with the double bonds and unbound electrons. If the sum is compatible with a value predicted by the formula, the compound is aromatic.

Compound 1 observes the rules for aromaticity. It has two π-electrons associated with the double bond. If n = 0, the formula would yield the integer 2. Thus, the compound is aromatic.

Compound 3 is not aromatic because this structure has four π-electrons, and the formula cannot give us a value of 4. The conjugate base of structure 3 is structure 2 (compound 2).

Compound 2 has a negative charge (two π-electrons) associated with it. These electrons occupy a p orbital and are delocalized within the ring. Compound 2 is aromatic because it has a total of six π-electrons associated with it. The formula can give a value of 6 if n = 1.

Compound 4 is aromatic. The lone pair on nitrogen lines up with the π-bonds of the ring and the system becomes aromatic. If n = 1, the formula predicts six electrons, because four π-electrons are associated with the two double bonds and there are two unpaired electrons.

Compound 5 is aromatic only if there is no donation of the lone pair into the ring. If you consider the lone pair in this structure, there would not be aromaticity.

Compound 6 is aromatic because of conjugation through the cyclic system and 10 π-electrons. Note that if n = 2, the formula predicts 10 π-electrons.

II. Key Reactions

A. ELECTROPHILIC AROMATIC ADDITIONS

Like alkenes and alkanes, the benzene ring is electron rich and the π-orbital electrons are accessible to electrophiles. Various electrophiles can add to benzene rings, but they are strongly elec-

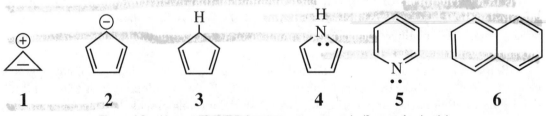

1   **2**   **3**   **4**   **5**   **6**

Figure 4-3. Various aromatic and nonaromatic compounds. (See text for details.)

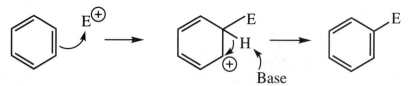

Figure 4-4. Mechanism of electrophilic aromatic addition.

trophilic and follow the same general reaction as shown in Figure 4-4. The electrophile is attacked by the ring and a resonance-stabilized carbocation forms. This intermediate is quickly deprotonated to re-establish aromaticity. An electrophilic aromatic addition product is formed.

Figure 4-5 depicts a particular example of electrophilic aromatic addition reaction: the **Friedel Crafts acylation reaction.** In this reaction, the $AlCl_3$ functions as the Lewis acid that polarizes the carbonyl (more than it is normally), and makes the carbon highly electrophilic. Remember that Lewis acids are good electron acceptors. The $AlCl_3$ pulls electrons from the Cl of the acid chloride, which weakens the C-Cl bond. Electrophilic addition results in the aromatic ketone.

Other groups can add to the ring to give addition products. The major reactions are **bromination, nitration,** and **alkylation.**

The important point here is that several different electrophilic aromatic additions can occur. Several examples are shown in Figure 4-6. Memorizing individual mechanisms is not necessary because all of these reactions are similar. Halogenation can occur as the diatomic halogen is polarized by the Lewis acid (in this case, $FeBr_3$). Nitration (adding an NO_2 group to the ring) can also occur. Chloroethane can be placed on a ring through Friedel Crafts alkylation. The Lewis acid activates the electrophile in a manner similar to that in the acylation described previously.

B. SUBSTITUENT EFFECTS

Whether a substituent takes electron density out of or donates electron density to the aromatic ring plays a large role in the physical characteristics and reactivity of the benzene ring. Remember that aromatic rings have fluffy electron clouds above and below the plane of the ring, and it

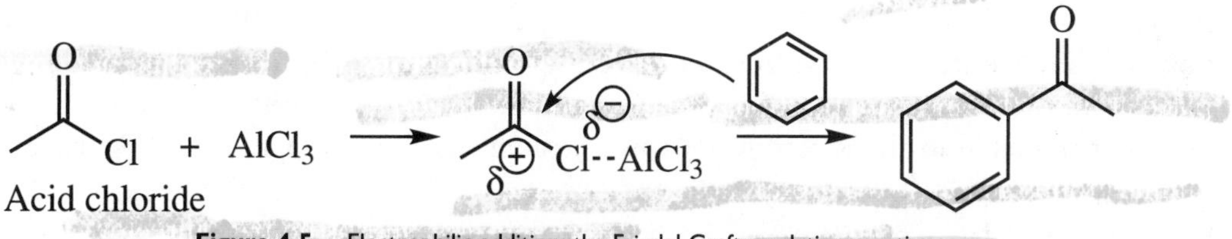

Figure 4-5. Electrophilic addition: the Friedel Crafts acylation reaction.

$Br_2 + FeBr_3 \longrightarrow Br_3Fe--Br-Br$ Bromination

$HNO_3, H_2SO_4 \longrightarrow NO_2$ Nitration

Cl $\xrightarrow{AlCl_3}$ $CH_3CH_2--Cl-AlCl_3$ Alkylation

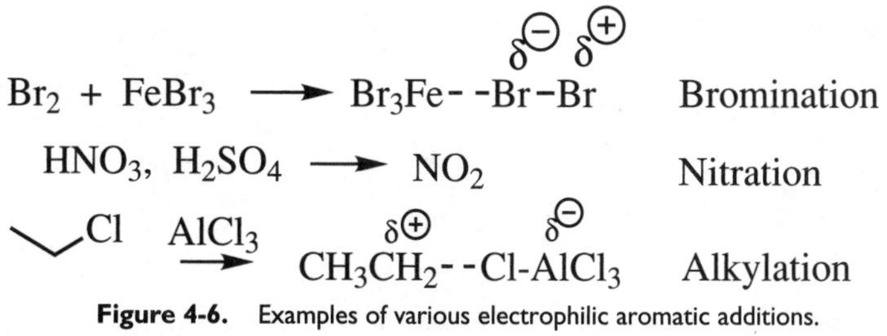

Figure 4-6. Examples of various electrophilic aromatic additions.

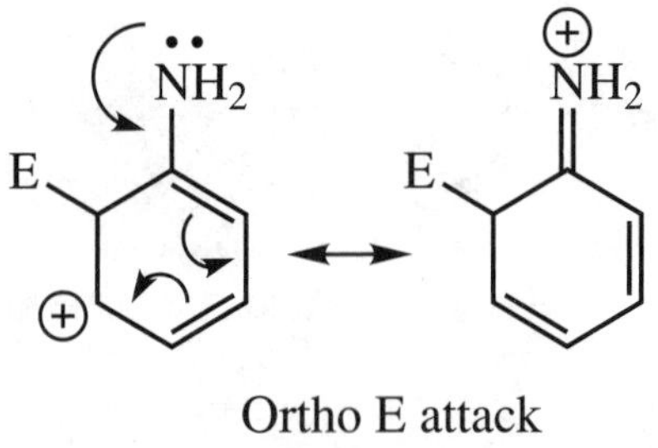

Ortho E attack

Figure 4-7. Activators: ortho/para directors. An example of how adding in the ortho position stabilizes the carbocation intermediate.

is these electrons that are susceptible to attack by an electrophile. If an electron-withdrawing group is placed on the ring, the relatively nonpolar benzene becomes polar because of the shift in the electron cloud toward the electronegative substituent. This change then alters the physical properties of the compound (e.g., melting and boiling points).

How is electrophilic addition affected if a group is already attached to the ring?

Any group attached to a benzene ring affects the **reactivity** of the ring and determines the **orientation** of substitution. **When an electrophilic reagent attacks an aromatic ring, it is the group already attached to the ring that determines where and how readily the attack occurs.**

1. ACTIVATORS AND DEACTIVATORS

A group that makes the ring more reactive than benzene is called an **activating group.** A group that makes the ring less reactive than benzene is a **deactivating group.**

a. Determination of Orientation

A group on a ring directs incoming substituents into an **ortho, para,** or **meta** position on the ring with respect to the initial group. For example, the -OH group of phenol directs an incoming substituent to an ortho or para position on the ring with respect to the -OH group. A nitro group on a benzene ring would direct incoming substituents to a meta position on the ring with respect to the nitro group.

b. Activators

Groups that are strong activators are ortho/para directors [i.e., electrophilic additions take place at the ortho and para positions (relative to the activator)]. For the most part, this orientation is attributable to the electron-releasing capability of the strongly activating substituents (amine groups and hydroxyl groups are good examples).

In Figure 4-7, the amino group is a strong activator and ortho/para director. The amino group directs to the ortho/para position because from these positions, the lone pairs on the nitrogen donate electrons into the ring to help stabilize the carbocation intermediate. This would not occur if substitution were at the meta position.

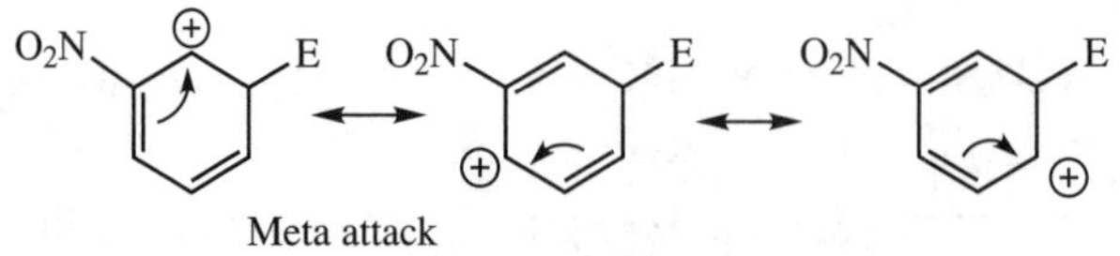

Meta attack

Figure 4-8. Deactivation: meta directors.

42

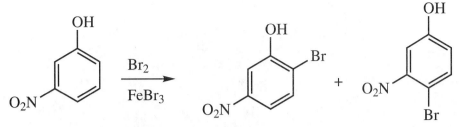

Figure 4-9. Example of priority assignment for deciding where electrophile will substitute on multisubstituted ring.

c. Deactivators

Deactivating substituents often direct to the meta position; halogens are an exception. Common deactivators and meta directors (Figure 4-8) include the nitro group, -CN group, -COOH group, and -SO₃H group. These substituents all have very δ^+ atoms attached directly to the ring and tend to draw electron density out of the ring, thus deactivating the ring toward electrophilic attack. Electrophilic addition can still occur, although when it does, these groups direct the incoming electrophile to the position meta to the substituent. Only after meta attack do you **not** get the highly unstable intermediate.

d. Halogens

Halogens are **deactivators** because of their electronegativity. They pull electron density out of the ring. Because of the presence of lone pairs, however, they are **also ortho/para directors.**

2. SUMMARY

Activating: *ortho/para directors* -NH₂, -NHR, -OH
Moderate activators, *ortho/para directors* -OCH₃
Weak activators, *ortho/para directors* -C₆H₅, -CH₃
Deactivating: *meta directors* -NO₂, -CN, -COOH, -SO₃, -CHO
Deactivating: *ortho/para directors* -F, -Cl, -Br, -I

If there is more than one substituent on the ring, refer to the following priority order (high to low) to decide where the electrophile will substitute:

o,p **activators** > *o,p* **deactivators (halogens)** > *m* **deactivators**

Figure 4-9 shows the o,p activating effects of the -OH group overriding the meta deactivating effects of the nitro group. Thus, Br adds ortho and para to the -OH group.

SECTION II

The Chemistry of Oxygen Containing Organic Compounds

Alcohols and Ethers $\quad$  5

I. Key Points

Alcohols and ethers contain an sp^3 hybridized oxygen that is electronegative and has two lone pairs of electrons. Alcohols and ethers differ in that **alcohols have a hydrogen bound to oxygen, whereas ethers have an alkyl group bound to oxygen.** This structural dissimilarity accounts for the noticeable difference in the chemistry between the compounds. (See the review of nomenclature for these compounds in the Appendix.)

II. Physical Properties

A. STRUCTURE

Because alcohols and ethers contain an sp^3 hybridized oxygen as the main functional group, the shape of the C-O-C bond is **bent,** with an angle of less than 109.5°. This configuration results from the repulsion of the lone pair of electrons of the oxygen (just as in water). The electronegativity of the oxygen polarizes the compound, which makes alcohols and ethers good solvents for polar organic compounds. Ethers are less reactive than alcohols because they lack the hydrogen bound to the oxygen.

B. MELTING AND BOILING POINTS

Alcohols can form intermolecular hydrogen bonds. Figure 5-1 shows the hydrogen bonding that can occur between two alcohol molecules. Note that the hydrogen of one molecule forms a hydrogen bond with the oxygen of another molecule. **Ethers cannot form intermolecular hydrogen bonds** because they have no hydrogen bound to oxygen. Thus, **alcohols have higher melting and boiling points than similar ethers.** The boiling point of ethers is similar to that of alkanes of comparable molecular weight.

An increase in chain length increases melting and boiling points for both alcohols and ethers, owing to an increase in the van der Waals interactions between molecules. Increased branching is associated with a decrease in van der Waals interactions, and a lowering of melting and boiling points.

Both short-chain alcohols and ethers can hydrogen bond with water, allowing them to mix with water. This ability indicates that the intermolecular interactions of water molecules can be replaced by interactions of the water with the alcohol or the ether. The water solubility of both alcohols and ethers decreases, however, with an increase in chain length. As the carbon chain length increases, the hydrophobic nature of these molecules increases.

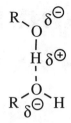

Figure 5-1. Hydrogen bonding of alcohols.

III. Reactions

Alcohols and ethers often undergo reactions in the presence of acids in which the oxygen becomes protonated. The positive charge is then reacted with a base or nucleophile.

A. REACTIONS WITH ACIDS: DEHYDRATION AND NUCLEOPHILIC ATTACK

1. ALCOHOLS

Treatment of alcohols with strong acids results in the displacement or elimination of the -OH group. Recall that -OH is a poor leaving group. Figure 5-2 illustrates the **substitution and elimination** reactions of alcohols. In step I, the oxygen donates a lone pair of electrons (Lewis base) and becomes protonated. In step II, water leaves to form a **carbocation;** this is the rate-limiting step. Finally, in step III, the carbocation is attacked by a nucleophile through an S_N1-type reaction or is eliminated by an E1 reaction.

Elimination occurs when the conjugate base of the acid is not a good nucleophile (as in the case of H_2SO_4). The carbocation formation occurs mainly with secondary and tertiary alcohols. Because of the nature of the reaction, rearrangements to form the more stable carbocation occur and result in a mixture of products. For primary alcohols, after protonation, water is displaced by a nucleophile in an S_N2-type reaction.

The S_N2-type reaction is also involved in the treatment of alcohols with PBr_3 or $SOCl_2$. These reactions result in the formation of alkyl bromides or chlorides, respectively (Figure 5-3). The O-P or O-S bond forms, putting a positive charge on the oxygen. Cl^- or Br^- acts as the nucleophile and displaces the species, resulting in the formation of the alkyl halide.

2. ETHERS

Similar reactions with acids occur with ethers by an S_N2 process. These reactions result in cleavage of the ethers (Figure 5-4). The acid protonates the oxygen, and this species is attacked by the conjugate base of the acid. In the example in Figure 5-4, the size of the nucleophile and bulk of

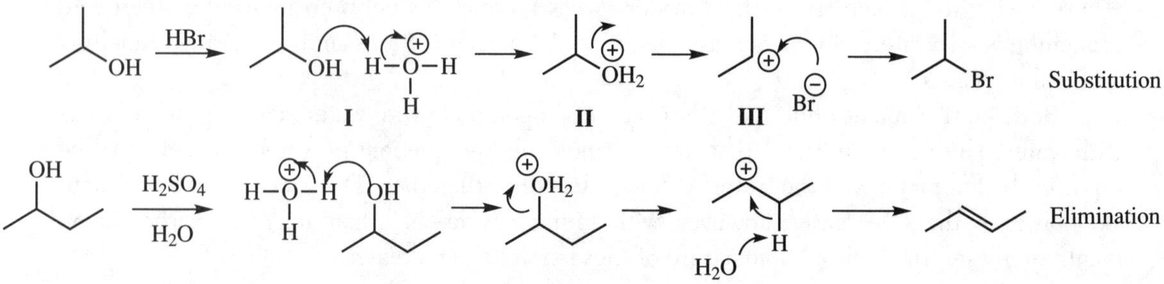

Figure 5-2. Substitution and elimination reactions of alcohols.

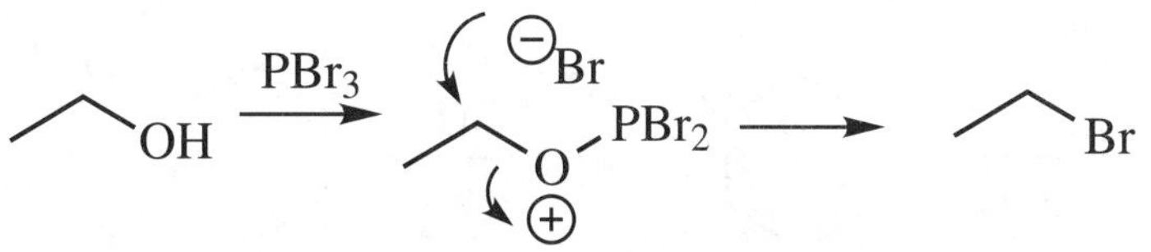

Figure 5-3. Formation of alkyl halides by reaction of alcohols with PBr$_3$ or SOCl$_2$.

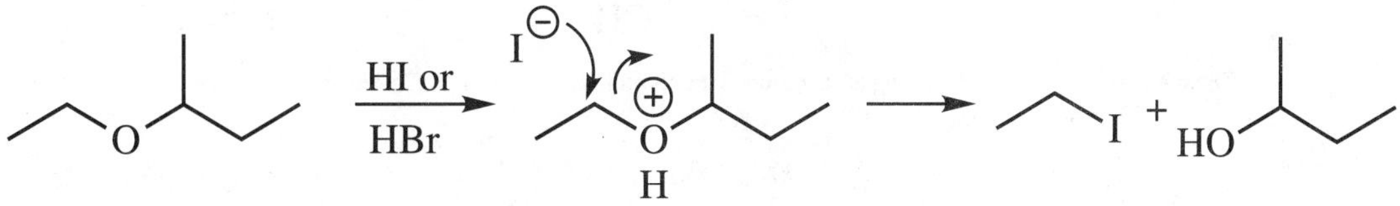

Figure 5-4. Cleavage of ethers by acid.

the electrophile cause the nucleophile to attack the primary electrophilic carbon instead of the more crowded carbon on the right. This reaction is an **anti-Markovnikov** addition of iodine.

B. ALCOHOLS AS NUCLEOPHILES

The C-O bonds of alcohols are polarized because of the electronegativity of oxygen. This polarization results in a δ^- at the oxygen, making it a nucleophile. **Alcohols can be used as solvent/nucleophiles for S$_N$1 reactions as a result of their polar protic nature.** (See examples of these reactions in Chapter 6.)

C. ETHER FORMATION

A popular way to synthesize an ether is to react an alcohol with a strong base. The resulting alkoxide is a highly nucleophilic species that reacts with primary alkyl halides and produces an ether by an S$_N$2 mechanism. This reaction is known as the **Williamson ether synthesis** (Figure 5-5).

D. EPOXIDES

Epoxides are three-membered, oxygen-containing rings that are highly strained owing to the acuteness of bond angles. They are susceptible to nucleophilic attack because of the electronegativity of oxygen, which causes polarization of the C-O bond.

Figure 5-6 demonstrates how epoxides are formed and how they may act as intermediates in reactions. In the reaction, a six-membered ring containing one double bond (halohydrin) is treated with base. An epoxide is formed (shown in the lower left structure of Figure 5-6). Note that the epoxide is in the form of a *cis* ring, which is opened. In this example, two enantiomers are formed, because the nucleophile can attack either carbon of the epoxide.

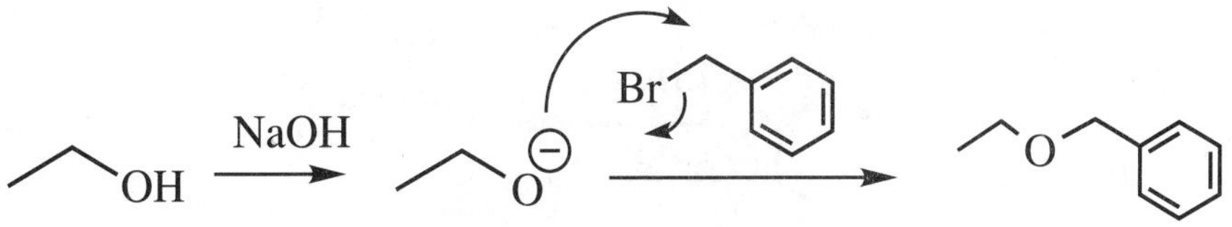

Figure 5-5. Williamson ether synthesis (synthesis of an ether from an alcohol).

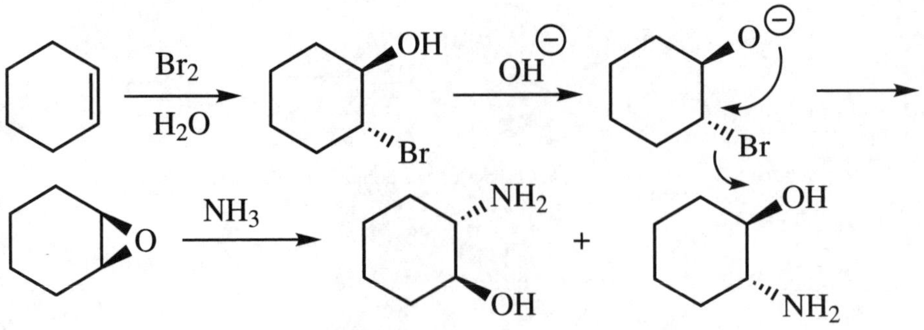

Figure 5-6. Reaction involving an epoxide. The epoxide is shown at the lower left of the figure.

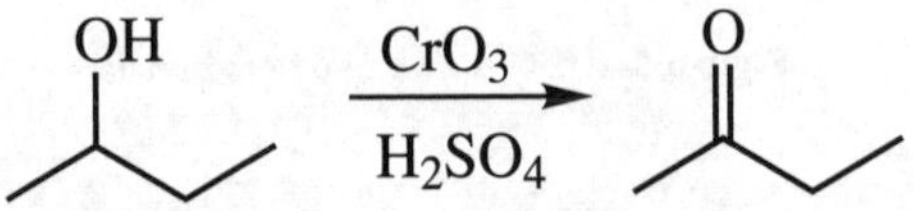

Figure 5-7. Oxidation of a secondary alcohol to a ketone.

E. OXIDATION OF ALCOHOLS

Chromium reagents, such as CrO_3, convert alcohols to carbonyl compounds. Secondary alcohols are oxidized to ketones (Figure 5-7).

Primary alcohols are converted to aldehydes and then to carboxylic acids (Figure 5-8). A special reagent, pyridinium chlorochromate, allows the isolation of the aldehyde that results from oxidation of the primary alcohol without overoxidation to the carboxylic acid.

F. FORMATION OF ALCOHOLS AND ETHERS BY REDUCTION

Many methods are used to synthesize alcohols; recall the hydration of alkenes and the hydroboration reaction discussed in Chapter 3. Another common method of generating alcohols is the reduction of carbonyl compounds. **Remember: you can recognize reducing agents by the large number of hydrogens.** Common reducing agents include $LiAlH_4$ and $NaBH_4$. Figure 5-9 illustrates the reduction of a ketone to an alcohol.

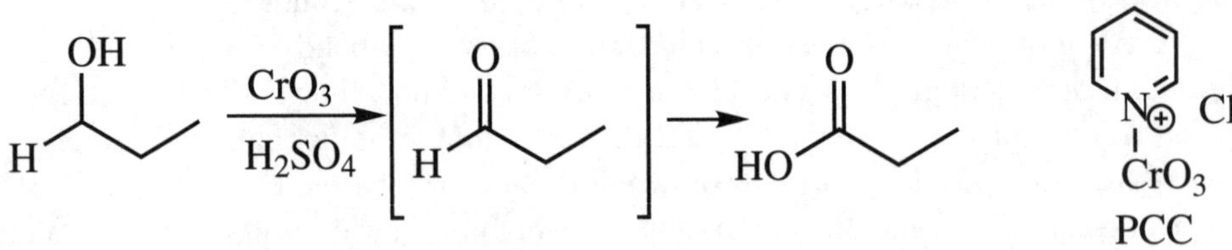

Figure 5-8. Oxidation of a primary alcohol to an aldehyde and ultimately to a carboxylic acid. The structure of the reagent pyridinium chlorochromate is also shown.

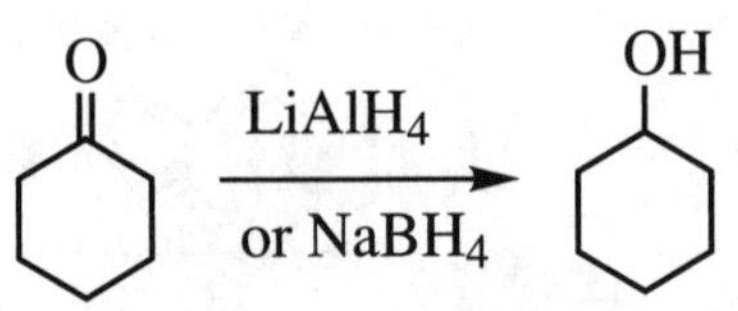

Figure 5-9. Formation of an alcohol by reduction.

50

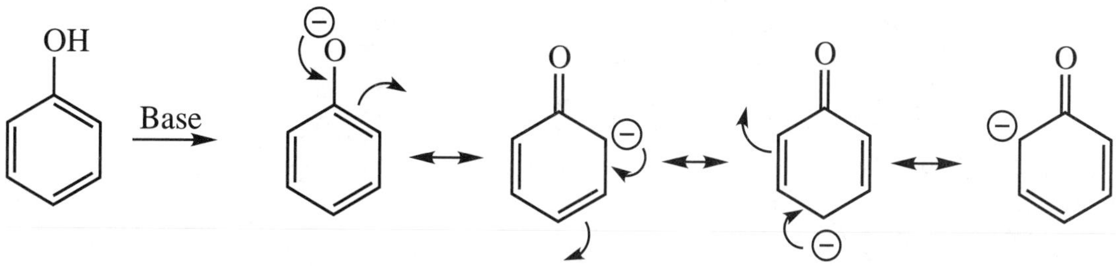

Figure 5-10. Resonance forms of phenol.

IV. Phenols

Discussion of phenols requires familiarity with benzene chemistry and acid–base theory. Phenols are special alcohols. They are **enols,** and favor the enol and not the ketone form because of aromatic stability. In addition, they are **more acidic than nonaromatic alcohols,** given the stability of the conjugate base formed on deprotonation. The resonance forms that stabilize phenol are shown in Figure 5-10.

Placement of electron-withdrawing groups on the ring enhances acidity. For example, phenol has a $pK_a = 9.89$, whereas the substituted phenols with electron-withdrawing nitro groups have lower pK_a values (m-NO_2 phenol: $pK_a = 8.28$; o-NO_2 phenol: $pK_a = 7.17$). The electron-withdrawing substituents enhance the stability of the conjugate base by the **inductive effect** (spreading the charge). The placement of these groups can also increase **resonance stability.** The anion can be stabilized through resonance with the electron-withdrawing group in the ortho and para positions (Figure 5-11). This situation does not occur when the group is in the meta position.

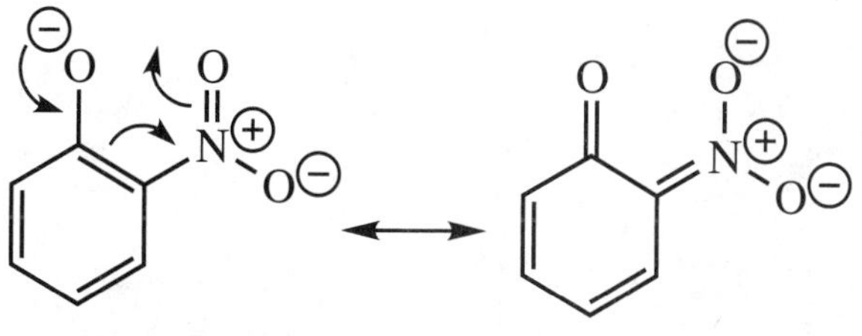

Figure 5-11. Resonance stability of an electron-withdrawing group in the ortho position. Substitution of an electron-withdrawing group increases acidity.

Carbonyl Compounds ⑥

The carbonyl compounds have many similar properties and reactions based on the chemistry of the carbonyl group that they share. Throughout the following review, consider the similarities and differences between the reactions of the aldehydes, ketones, and carboxylic acids and their derivatives based on the carbonyl group and the structural differences between the molecules. (See the review of the nomenclature for these compounds in the Appendix.)

I. Physical Characteristics and Properties

Carbonyl compounds have a π-bond and a σ-bond formed between the carbon and the oxygen. The chemistry of these compounds is dictated by their structure and by the dipole between the electronegative oxygen and the partially positive sp^2 carbon. Figure 6-1 illustrates the bonds and resonance forms of the carbonyl group. Note that the carbonyl carbon is electrophilic and the carbonyl oxygen is nucleophilic.

A major set of the reactions of the carbonyl compounds involves the nucleophilic attack of the carbonyl group electrophilic carbon. The consequences of nucleophilic attack depend on the type of carbonyl compound involved (with or without a good leaving group).

A. STRUCTURE

The carbonyl carbon is sp^2 hybridized, which makes the carbonyl structure **planar** with bond angles of 120°. The electrophilic nature of the carbonyl carbon is clarified in Figure 6-1. Note that the resonance form for the carbonyl group has a positive charge on the carbon, which renders the carbon susceptible to nucleophilic attack.

B. MELTING AND BOILING POINTS

Carbonyl compounds have higher boiling points than comparable alkanes. This difference relates to the enhanced **dipole–dipole interactions** between molecules with carbonyl groups.

Ketones lack intermolecular hydrogen bonding because they do not have hydrogens associated with oxygen. Thus ketones have lower boiling and melting points than comparable alcohols.

Hydrogen bonding does occur for carboxylic acids. These hydrogen bonds must be overcome for boiling to occur. Thus, **carboxylic acids have higher melting and boiling points than aldehydes or ketones.** In addition, because the carboxylic acids have strong dipole–dipole interactions, they **have higher melting and boiling points than alcohols of comparable length.**

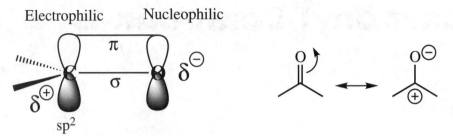

Figure 6-1. Bonds and resonance forms of the carbonyl group.

II. Important Implications of the Carbonyl Group

A. KETO-ENOL TAUTOMERISM

Carbonyl compounds with hydrogens on the α-carbon (carbon next to the carbonyl carbon) are rapidly interconvertible between a keto form and an enol form. Recall that an enol is a structure with an -OH group attached to a doubly bonded carbon. The *-ene* suffix is indicative of a carbon–carbon double bond, and the *-ol* is used for the alcohol group. The keto form contains a carbonyl group.

The rapid interconversion that occurs between the keto form and the enol form is referred to as **tautomerism.** This interconversion occurs between compounds with structures that differ greatly in the arrangement of atoms but exist in easy and rapid equilibrium. As is true of keto–enol tautomerism, most tautomerisms involve structures that differ in the point of attachment of hydrogen. Typically, the equilibrium of this interconversion overwhelmingly favors the structure in which hydrogen is bonded to carbon (keto form). **The keto form is also more stable than the enol form.**

Figure 6-2 illustrates **keto–enol tautomerism,** a process that can be catalyzed by base or acid. Note that whether catalyzed in base or acid, the process starts with the carbonyl group in the form of a ketone. A hydrogen is extracted from the α-carbon, and intermediates form. The result is a structure with a double bond and a hydroxyl group—the enol form. **Remember: the keto form is more stable than the enol form, and rapid interconversions between these two forms occur.**

B. CARBOXYLIC ACIDS

These compounds are **acidic because of the delocalization of the electrons** of the conjugate base (carboxylate ion) over the two oxygens and the carbonyl carbon. Recall that inductive effects can enhance acidity by further polarizing the O-H bond and stabilizing the conjugate base.

If a group is close enough or is connected through an electron cloud, an electronegative moiety will enhance acidity of the carboxylic acid by helping to stabilize the conjugate base. In

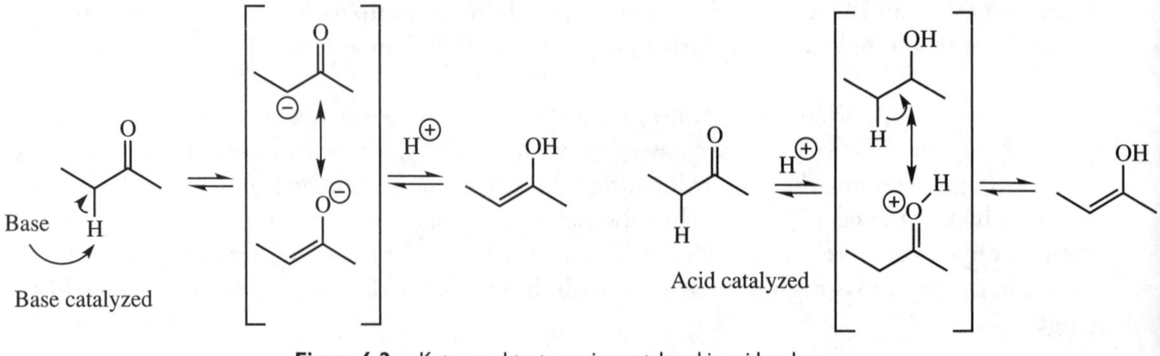

Figure 6-2. Keto–enol tautomerism catalyzed in acid or base.

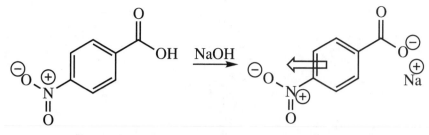

Figure 6-3. How an electron-withdrawing group increases acidity.

Figure 6-3, the electron-withdrawing nitro group removes electron density from the ring and stabilizes the anion formed after deprotonation.

When comparing the acidity of carboxylic acids with electron-withdrawing halides attached near the carbonyl carbon, remember these general rules:

1. **The more halides attached to the acid molecule, the stronger the acid.**
2. **The closer the halide atoms are attached to the carboxylic acid functional group, the stronger the acid.**

C. AMIDES

These compounds also have a carbonyl group. **Amides are considered functional derivatives of carboxylic acids because instead of the -OH group of the acid, they have a -NH₂ group.** Like the carboxylic acids, they have **high melting and boiling points because of hydrogen bond formation.** Amides have the ability to form intermolecular hydrogen bonds as long as there is a hydrogen on nitrogen. Figure 6-4 shows the intermolecular hydrogen bonding possible with amides and the resonance forms associated with amides.

These compounds are also unique in the ability of the nitrogen to donate electrons and form a π-bond with the carbonyl carbon. This electron donation stabilizes the resonance hybrid. The C-N bond is said to have partial double-bond character. Thus, it does not rotate as easily as a normal C-N single bond. This characteristic has important implications in the structure of peptides and proteins with amide bonds between the individual amino acids.

D. ACIDITY OF α-PROTONS IN CARBONYL COMPOUNDS AND THE ENOLATE ION

The protons α to the carbonyl (except for carboxylic acids and amides) are slightly acidic. The pK$_a$ of the proton in this position varies from 17 to 30, depending on the type of carbonyl compound. The anion at this position, the **enolate ion,** is stabilized through resonance, allowing for delocalization of the negative charge.

The formation of the enolate ion (Figure 6-5) should look familiar; compare it to the keto–enol interconversion reviewed previously (see Figure 6-2). The enolate ion is critical in

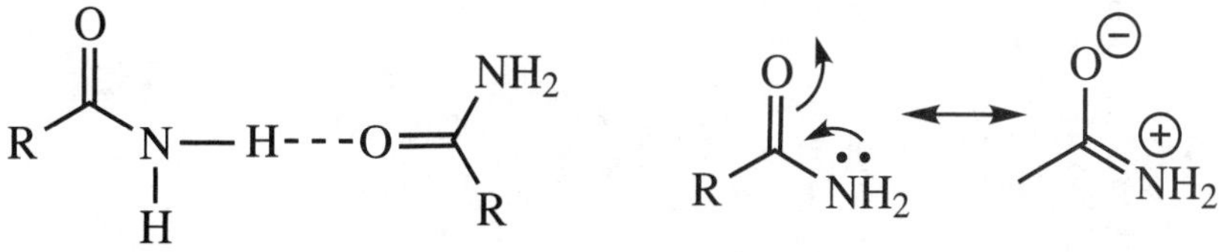

Figure 6-4. Hydrogen bonding and resonance structures associated with amides.

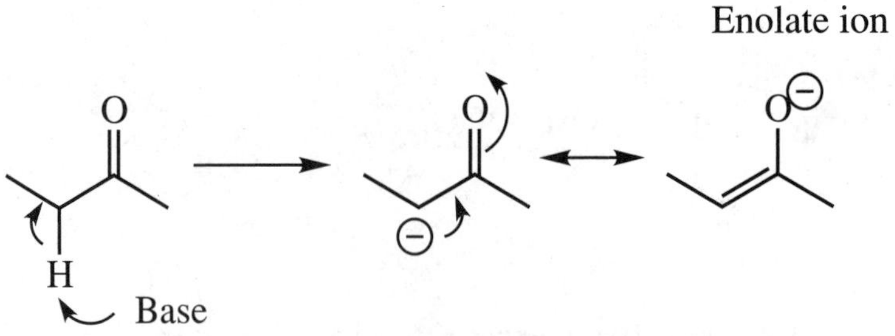

Enolate ion

Figure 6-5. Formation of the enolate ion by extraction of proton α to the carbonyl.

organic chemistry because it is a **good nucleophile.** The importance of this ion in chemical reactions is discussed subsequently.

III. Key Reactions

A. OVERVIEW

The basic reactions of carbonyl compounds involve **nucleophilic additions and substitutions, reductions, and reactions involving enolates.**

Figure 6-6 illustrates the paths and possible results of nucleophilic attack: **nucleophilic addition (aldehydes and ketones)** and **nucleophilic substitution (carboxylic acids).** Direct addition and conversion of the carbonyl to another species is shown along **path A.** Displacement of a leaving group is shown along **path B.** Both paths go through a tetrahedral (sp^3) intermediate. The nucleophile can attack from either face of the sp^2 hybridized carbon and the π-electrons shift to the oxygen, giving it a negative charge.

By understanding that **nucleophilic attack of aldehydes and ketones tends to follow a nucleophilic addition reaction (path A)** and **nucleophilic attack of carboxylic acids tends to follow a nucleophilic substitution reaction (path B),** you can predict the products of many reactions.

Reduction (addition of H$_2$ across a double bond) can also occur, yielding aldehydes (under careful control) and alcohols (Figure 6-7).

Another main type of reaction is based on the **reactivity of enolates.** These ions are compounds formed from the removal of the proton α to the carbonyl. Enolates are good nucleophiles that can undergo a variety of reactions with various electrophiles.

B. NUCLEOPHILIC ADDITION TO ALDEHYDES AND KETONES

Aldehydes and ketones possess very poor leaving groups adjacent to the carbonyl, which dictates the outcome of their reactions.

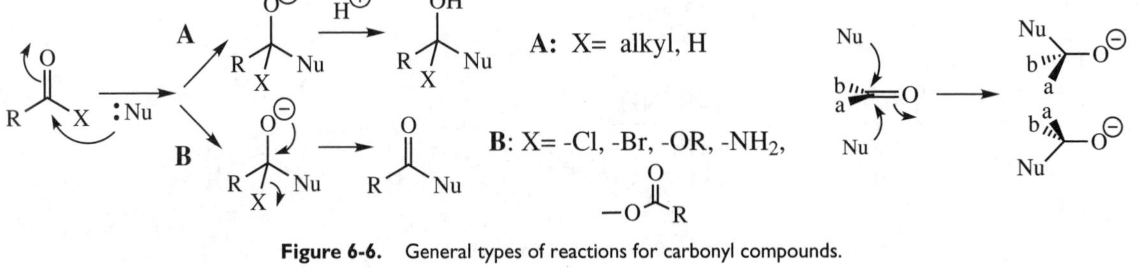

Figure 6-6. General types of reactions for carbonyl compounds.

56

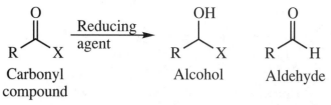

Figure 6-7. Reduction of carbonyls to yield alcohols and aldehydes.

1.   ADDITION OF STRONG NUCLEOPHILES

Nucleophiles can add to the carbon and form a wide variety of alcohols, a process that follows the mechanism of nucleophilic addition shown in Figure 6-6. Nucleophiles, such as Grignards (RMgBr), organocuprates (R_2CuLi), and acetylides (Figure 6-8), are needed for these reactions to occur. Addition is favored because of the poor leaving groups of the aldehydes and ketones. In each example in Figure 6-8, note: (1) that a ketone is attacked by a strong nucleophile and an alcohol is formed, and (2) that a group has added to the carbonyl carbon.

2.   HEMIACETAL/HEMIKETAL AND ACETAL/KETAL FORMATION

These reactions involve the attack of the electrophilic carbon by an alcohol. When a single alcohol molecule undergoes nucleophilic addition and adds to the carbonyl group, a **hemiacetal** (if starting compound is an aldehyde) or a **hemiketal** (if starting compound is a ketone) forms. The addition of a second alcohol molecule by nucleophilic addition gives **the acetal or ketal.** These reactions can be catalyzed by acid or base.

Study the reaction mechanism in Figure 6-9. Note that the acid protonates the carbonyl of the ketone and the alcohol attacks the electrophilic carbon, leading to the production of the hemiketal. In general, hemiketals and hemiacetals are too unstable to be isolated. The reaction can continue with transfer of a proton to the -OH group and elimination of water. The carbon is then attacked by another alcohol molecule to form the ketal. If the starting compound is an aldehyde, a hemiacetal and an acetal form according to the reaction mechanism just described.

Reaction arrows indicate that the reactions are readily reversible. To favor one side over the other, the conditions must be carefully controlled. Excess alcohol shifts the reactions to the right.

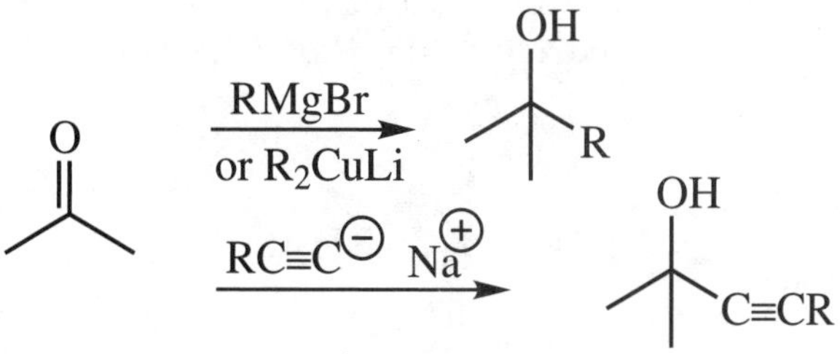

Figure 6-8. Various nucleophilic addition reactions.

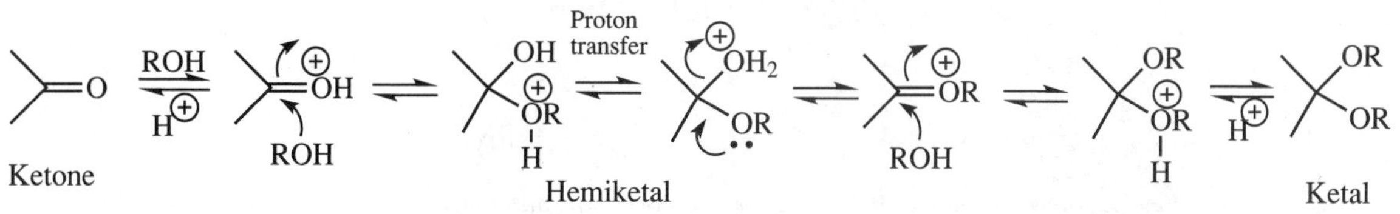

Figure 6-9. Hemiketal and ketal formation.

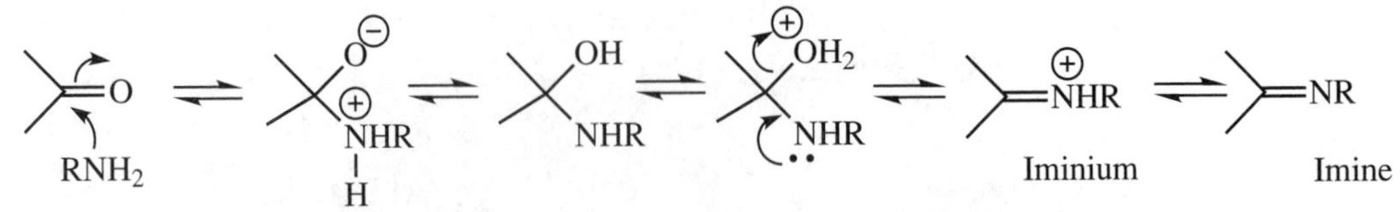

Figure 6-10. Addition of amine to a carbonyl yields an imine.

The addition of water and acid shifts the reactions to the left and restores the carbonyl. The acid is a catalyst and is neither consumed nor generated.

3. ADDITION OF AMINES

Aldehydes and ketones can be attacked by amines, resulting in imines and iminium ions (Figure 6-10). **An imine is a compound that contains a carbon–nitrogen double bond.** Note how these reactions differ from the classic nucleophilic addition reactions discussed previously. After the initial attack of the amine, a proton is transferred and water is displaced in a manner similar to acetal or ketal formation.

C. NUCLEOPHILIC SUBSTITUTION TO CARBOXYLIC ACID DERIVATIVES

Various products of nucleophilic substitution reactions are illustrated in Figure 6-11; a range of nucleophiles reacts readily with these compounds. **The products of nucleophilic substitution reactions include carboxylic acids, acid chlorides, amides, esters, ketones, and anhydrides.** All of the reactions go through a **tetrahedral intermediate.** The leaving group is then expelled. The reactivity of the carbonyl is directly related to the quality of X as a leaving group. Remember from $S_N1,2$ that the more stable anions are better leaving groups. The reactivity order of leaving groups, from high to low, is shown in Figure 6-12.

As demonstrated in Figure 6-11, carboxylic acid derivatives can be synthesized through nucleophilic substitution. These reactions occur most readily from **acid chlorides,** which can be synthesized from the reaction of the carboxylic acid with $SOCl_2$ or PCl_3, using a mechanism similar to that for generating alkyl halides from alcohols (see Chapter 5 III A).

D. REDUCTION OF CARBONYL COMPOUNDS

Carbonyl compounds are reduced by agents that deliver hydrides (H^-), such as **lithium aluminum hydride (LiAlH$_4$), sodium borohydride (NaBH$_4$), borane (BH$_3$) and their deriva-**

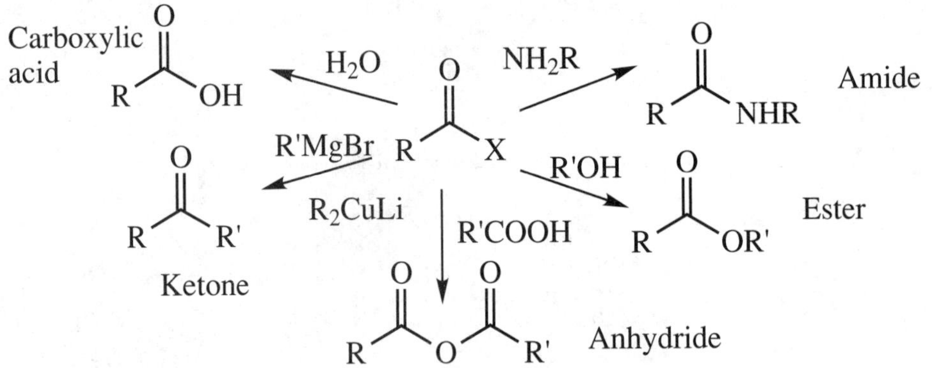

Figure 6-11. Various products of nucleophilic substitution reactions. Note the production of various carboxylic acid derivatives.

58

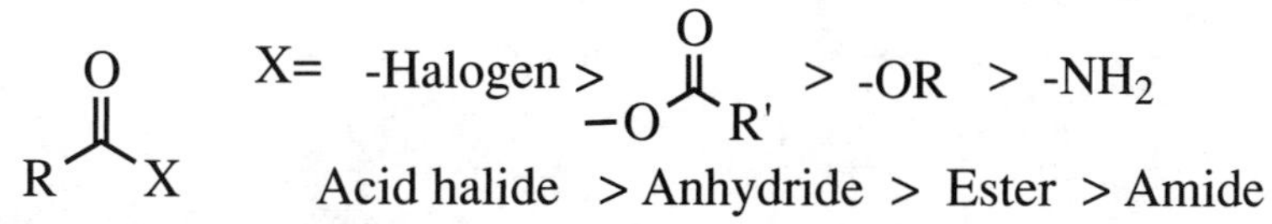

X= -Halogen > $\underset{-O}{\overset{O}{\|}}$ R' > -OR > -NH$_2$

Acid halide > Anhydride > Ester > Amide

Figure 6-12. Leaving group strength in nucleophilic substitution reactions.

tives. These reagents have certain selectivities, with the exception of **LiAlH$_4$, which is the strongest reducing agent** and reduces acids, aldehydes, and ketones down to an alcohol. LiAlH$_4$ also reduces amides and nitriles to amines. Ketones reduce to secondary alcohols and aldehydes, and carboxylic acid derivatives reduce to primary alcohols. The more sterically hindered agents [such as LiAlH(OtBu)$_3$] are not able to deliver the hydride as effectively as LiAlH$_4$, and thus are considered milder reducing agents. Consequently, they are able to reduce carboxylic acid derivatives to aldehydes without further reduction to the alcohol (Figure 6-13).

Catalytic hydrogenation can also be used to reduce carbonyl compounds to alcohols (from ketones, aldehydes, esters, and the like) and amines (from amides, nitriles). Heterogeneous catalysts such as Ni and Pt are often used. Note that any double bonds contained in the carbonyl compounds are reduced as well.

E. FORMATION OF CARBONYL COMPOUNDS

Aldehydes and ketones can be synthesized by various methods discussed previously, including the **oxidation of alcohols** (see Chapter 5 III E) and the **oxidation of alkenes** (ozonolysis; see Chapter 3 II D).

Oxidation of alcohols and aldehydes by the strong oxidant CrO$_3$ yields carboxylic acids.

The Grignard synthesis reaction is another way to synthesize carboxylic acids. The **Grignard reagent,** one of the most useful and versatile reagents known in organic chemistry, **has the general formula RMgX and the general name alkylmagnesium halide.** Basically, it is prepared by adding MgX to an alkyl group.

When the Grignard reagent is reacted with carbon dioxide, the alkyl group adds to the carbon–oxygen double bond of carbon dioxide. In the presence of mineral acid, the carboxylic acid is synthesized. Figure 6-14 shows how a Grignard reagent containing a phenyl group reacts with carbon dioxide in the presence of acid. The phenyl Grignard acts as a nucleophile and attacks carbon dioxide at the electrophilic carbon to form the carboxylic acid. This reaction is performed by pouring the Grignard reagent on crushed dry ice (carbon dioxide), or by bubbling carbon dioxide gas into an ether solution of the Grignard reagent.

Figure 6-14 also demonstrates that **carboxylic acids can be synthesized by the hydrolysis of nitriles.** Nitriles contain an R group and a carbon atom triple bonded to a nitrogen atom **(RC≡N),** and are formed by treatment of alkyl halides with sodium cyanide in a nucleophilic substitution reaction. In the presence of water and acid, the nitrile is hydrolyzed to the carboxylic acid.

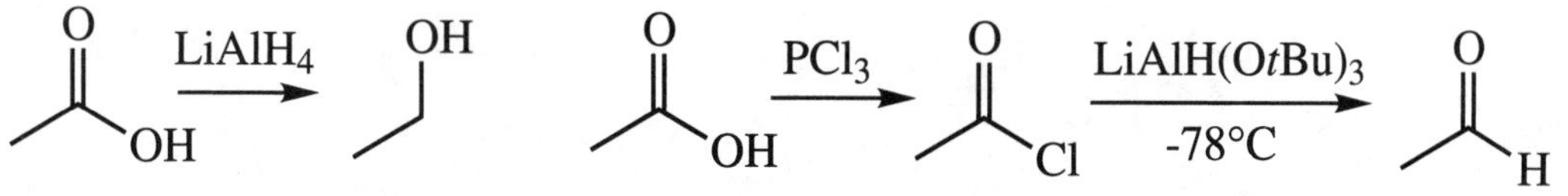

Figure 6-13. Reduction of carboxylic acid to alcohol by LiAlH$_4$ (right), and conversion of a carboxylic acid to an acid chloride followed by reduction to an aldehyde by a mild reducing agent (left).

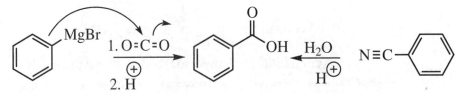

Figure 6-14. Synthesis of carboxylic acids by Grignard reaction (left) and nitrile synthesis (right).

Esters are formed through acid-catalyzed attack of the electrophilic carbonyl carbon by an alcohol. Figure 6-15 shows the production of an ester by reaction of a carboxylic acid and an alcohol. In practical terms, it is easier to form esters from carboxylic acid derivatives, such as acid chlorides.

F. REACTIONS OF ENOLATES

Recall that enolate ions are good nucleophiles and are important in the chemistry of carbonyl compounds. Enolate ions are often made by reacting a carbonyl compound with a base that extracts the α-proton.

1. HALOGENATION AND ALKYLATION

Reactions of enolates lead to **halogenation and alkylation** of carbonyl compounds. After enolate formation, halogenation of the α-position occurs as the carbon anion attacks the halogen (Figure 6-16). Without careful control, a second halogen may add to the α-position because of the acidic α-proton that is present in the monohalogenated product.

The enolate is a good nucleophile that can displace halides from alkyl halides to yield alkylated carbonyl compounds (see Figure 6-16, at right). Note that as the steric hindrance about the alkyl halide increases, the chance of E2 competing with the S_N2 alkylation increases.

2. REACTIONS WITH ALDEHYDES AND KETONES (ALDOL CONDENSATION)

Under the influence of dilute acid or base, **two molecules of aldehyde or ketone may combine. The products of this combination are called β-hydroxyaldehydes or β-hydroxyketones.** This reaction is called the **aldol condensation** (Figure 6-17).

This reaction occurs as follows. Remember that enolate ions are effective nucleophiles and can attack other carbonyls. As observed in other reactions, the nucleophile attacks the electrophilic carbon and forms a tetrahedral intermediate. This intermediate is a β-hydroxy carbonyl compound, known as an aldol or ketol. After heating, dehydration occurs (with acid or base catalysis) to form the more stable conjugated α,β-unsaturated carbonyl system. Deprotonation occurs because the α-proton is still acidic. Note that two carbonyl molecules have combined.

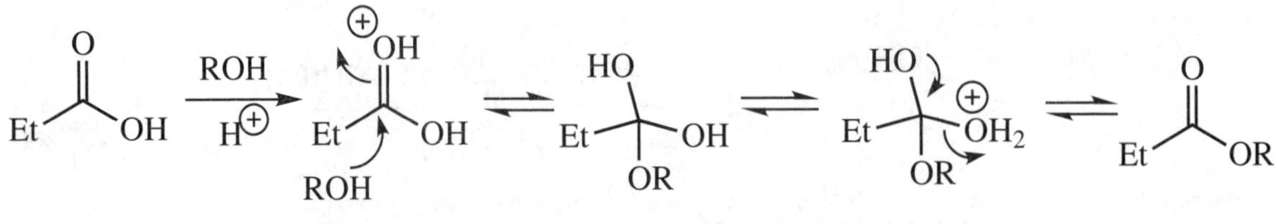

Figure 6-15. Formation of esters from carboxylic acids and alcohols.

SECTION II • CARBONYL COMPOUNDS

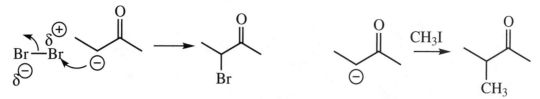

Figure 6-16. Halogenation and alkylation reactions involving enolates.

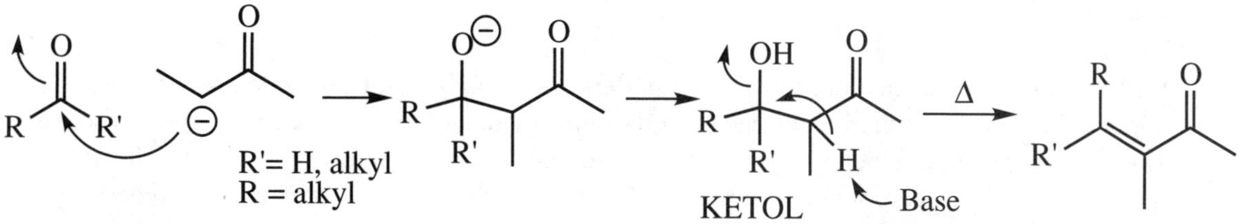

R'= H, alkyl
R = alkyl

R'

KETOL   ⤺ Base

R

R'

Figure 6-17. Aldol condensation. Note enolate attack of carbonyl, formation of intermediate, and dehydration, yielding product.

In every case, the product of this reaction results from the addition of one molecule of aldehyde or ketone to a second molecule in such a way that the α-carbon of the first becomes attached to the carbonyl carbon of the second. If both compounds contain α-protons, the result may be a mixture of products. To provide more control over product formation, this reaction sometimes is carried out with one reactant that does not have α-protons.

3. β-KETO ESTERS

a. Formation

Enolates, being good nucleophiles, are able to react with carboxylic acid derivatives. **β-Keto esters are formed from the condensation of two esters:** one acting as the enolate and the other acting as the electrophile (Figure 6-18). The initial step of this reaction yields an intermediate that has an α-proton to two carbonyl groups. A base (ethoxide) is used to deprotonate this proton. In the example in Figure 6-18, the product of the reaction can then be stabilized by resonance forms.

b. Decarboxylation

The reaction depicted in Figure 6-19 starts with an ester being hydrolyzed with base, forming the β-keto acid. Remember that a β-keto acid has a keto group in a position β to the carboxyl group.

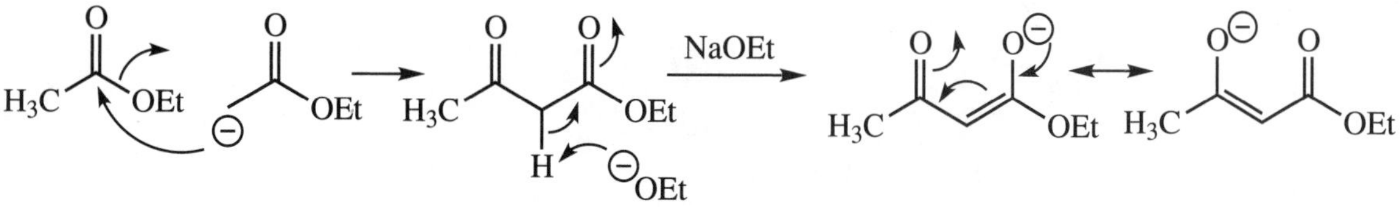

H_3C OEt   OEt H_3C OEt   NaOEt H_3C OEt H_3C OEt

H   OEt

Figure 6-18. Formation of β-keto esters by condensation of two ester molecules.

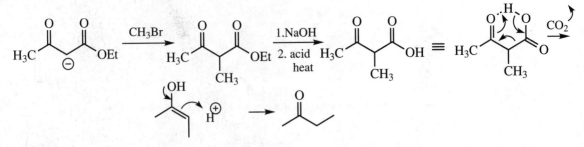

Figure 6-19. Decarboxylation reaction. An ester is hydrolyzed to a β-keto acid before being decarboxylated. The final product is a ketone.

With protonation and heating, this species undergoes **decarboxylation,** which occurs as the six-membered, intramolecularly hydrogen-bonded intermediate forms between the two carbonyls. The enol that results quickly converts to the ketone. This reaction occurs for any dicarbonyl species in which the carbonyls are separated by an sp^3 carbon and one is a protonated carboxylic acid.

Organic Molecules of Biologic Importance and Separation/ Purification of Organic Compounds

org

anic

chemi

......

Carbohydrates $\bigcirc 7$

From a biologic standpoint, the bonds of carbohydrates are nature's way of storing solar energy. Metabolism results in the oxidation of this stored energy and the synthesis of adenosine triphosphate and other high-energy molecules. Carbohydrates, commonly called sugars, are also involved in other biologically important molecules, such as nucleic acids (DNA and RNA), and in polymeric forms, such as cellulose (cell walls) and glycogen (storage form of glucose in the liver).

I. Structure and Physical Properties

Think of simple carbohydrates as polyhydroxylated aldehydes and ketones (see Chapters 5 and 6 of the Organic Chemistry Review Notes). The simplest carbohydrates are the monosaccharides (Figure 7-1). If the sugar contains an aldehyde group, it is an aldose. If it contains a keto group, it is a ketose. All the monosaccharides shown in Figure 7-1 are aldoses. A ketose is shown in Figure 7-2.

By convention, the monosaccharide carbon backbone is numbered, starting from the end closest to the most highly oxidized carbon. In Figure 7-1, carbon 1 of glucose would be the carbon with the aldehyde group at the top of the molecule. Carbon 6 would be the carbon at the bottom of the molecule (CH_2OH).

A monosaccharide is identified on the basis of the number of carbons it contains (i.e., triose, tetrose, pentose, hexose, and so on). The most common monosaccharides in our diet are hexoses (glucose and fructose).

Monosaccharides are often represented as Fischer diagrams, and are designated D or L. These designations should not be confused with d and l, which designate the direction of the rotation of plane-polarized light. The D and L designations represent the relationship of a carbohydrate to the structure of D- or L-glyceraldehyde.

The compound glyceraldehyde (see Figure 7-1) was selected as the standard of reference for which the D and L configurations are named. Compounds related to D-glyceraldehyde are designated D, and the compounds related to L-glyceraldehyde are designated L. All monosaccharides are named D or L on the basis of the lowest chiral center, the carbonyl group being at the top. Thus, if -OH is on the right of the lowest chiral center, the sugar has a D-designation. The L -designation indicates that the chiral carbon furthest from the carbonyl has a hydroxy group on the left.

In Figure 7-1, note that the D-glucose has its -OH group on the right side of carbon 5 (lowest chiral center for glucose), and L-glucose has this -OH group on the left side. Check mannose and galactose to see if you agree with the designations given.

By recalling the concepts of enantiomers and diastereomers (see Chapter 2), you can determine which of the hexoses in Figure 7-1 are enantiomers and which are diastereomers: D- and

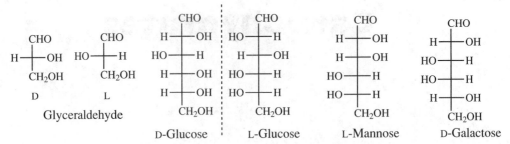

Figure 7-1. Configuration of D- and L-glyceraldehyde and several aldohexoses.

L-glucose are enantiomers, whereas L-glucose and D-galactose are diastereomers. L-glucose and L-mannose are also diastereomers. Note that enantiomers are D or L with the same name; diastereomers have different names.

Remember:

Epimers are diastereomers that differ in the chirality about carbon 2. L-glucose and L -mannose are epimers because they differ in the chirality of carbon 2.

Anomers are diastereomers that differ in configuration about carbon 1. Anomeric carbons are reviewed subsequently.

A. ENOLIZATION TO KETOSES

Aldoses can enolize to yield ketoses (see Figure 7-2). This process involves the removal of an α-proton by a base, leading to the enolate, and protonation to give the enediol. Tautomerization then gives the ketose. Treatment of the ketose with base reverses the reaction. The reverse reaction yields two products because of racemization of the carbon 2 position; protonation can occur on either face of the planar enolate when regenerating the aldose.

B. CYCLIZATION

Sugars in solution exist in equilibrium with linear and cyclic forms; this is particularly the case with pentoses and hexoses. Cyclization leads to the formation of hemiacetals. **Note:** This cyclization step is the same reaction as described in the discussion of hemiacetal formation in Chapter 5.

In Figure 7-3, note that the free -OH group on carbon 5 can attack the carbonyl carbon, which has its oxygen in two different orientations. This interaction gives rise to the two different cyclic hemiacetals, an α- and a β-form. These forms are **anomers** (diastereomers that differ in configuration about carbon 1), and the carbon with the different stereochemistry is the **anomeric carbon.** Structurally, whether the α- or β-anomer results depends on the orientation of the aldehyde as it is attacked by the hydroxy group of the carbon.

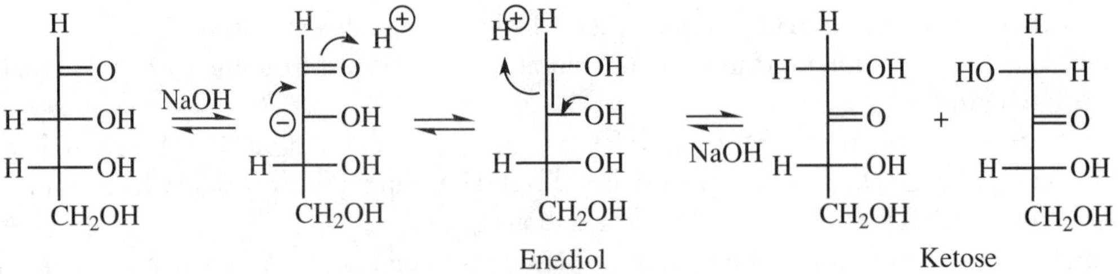

Figure 7-2. Conversion of aldoses to ketoses by enolization.

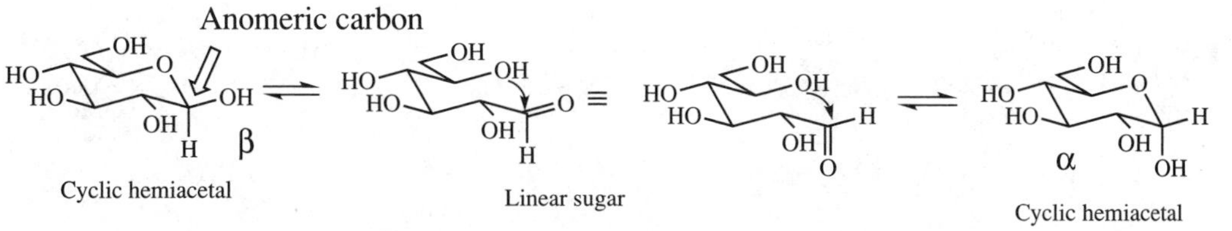

Anomeric carbon

Figure 7-3. Conversion of linear forms to cyclic α- and β-hemiacetal forms. Note the involvement of the anomeric carbon in generating α- and β-forms.

D-glucose exists in equilibrium with two cyclic forms (α and β) and the linear aldose form. At any given moment, most of the molecules are in the cyclic form; between the two cyclic forms, the β-form predominates (64% β versus 36% α). In glucose, **the β-form predominates because it is more stable.** The -OH group attached to carbon 1 of the β-anomer is in an equatorial position on the chair, which is preferable to the **axial position of the -OH group of the α-anomer.** If you put 100% α-glucose (solid) in solution, its optical rotation starts to change as the α and β equilibrate. This effect is called **mutarotation.**

II. Key Reactions

A. GLYCOSIDE FORMATION

Glycosides are acetals that, just like normal aldehydes, are generated from the reaction of the aldehyde with an alcohol and acid (Figure 7-4). The reaction starts when an acid protonates the hydroxy group of the hemiacetal. After water leaves, a resonance-stabilized carbocation forms, the only possible carbocation that is resonance stabilized by the oxygen. The nucleophilic alcohol then attacks the carbocation from the bottom or top face, giving both α- and β-forms. Once formed, the glycoside **does not undergo mutarotation under normal conditions,** because the compound is no longer a hemiacetal.

Glycosides are also referred to as **nonreducing sugars.** If these compounds are tested with Tollen's reagent (which identifies aldehydes by reduction of a silver salt to silver metal), they show a negative ("nonreducing") result. **In contrast, all monosaccharides are reducing sugars** and reduce Tollen's reagent.

Glycosides are stable to base, but when exposed to acid and water, they revert back to the sugar. The equilibrium shifts to the left, the sugar is produced, and the -OR group is released. This process is the same by which polysaccharide linkages at the carbon 1 position are hydrolyzed by acid and water. The sugar then falls back into the equilibrium of α-, linear, and β-forms.

B. ETHER FORMATION

It is easy to methylate the hydroxyl groups of a glycoside by a reaction similar to the Williamson synthesis reaction (Chapter 5 III C) called **exhaustive methylation.** The formation of ether

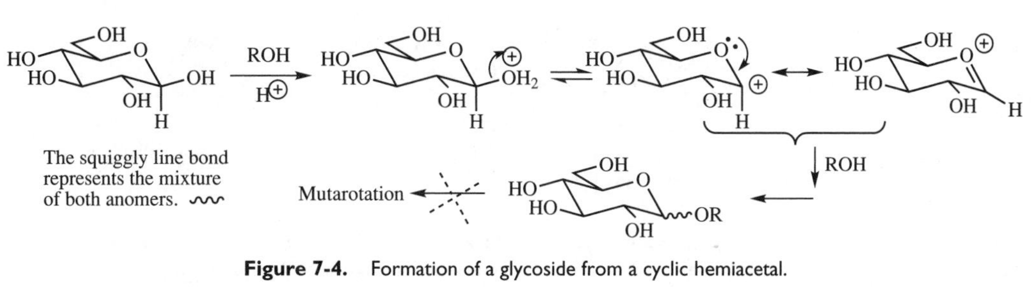

The squiggly line bond represents the mixture of both anomers. ∿

Mutarotation

Figure 7-4. Formation of a glycoside from a cyclic hemiacetal.

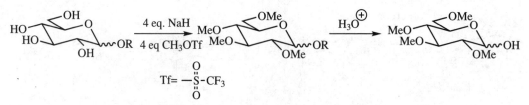

$$\text{Tf} = -\overset{\overset{\displaystyle O}{\|}}{\underset{\underset{\displaystyle O}{\|}}{S}}-CF_3$$

Figure 7-5. Ether formation, an example of exhaustive methylation of a glycoside.

(Figure 7-5) is an example of this process. Exhaustive methylation refers to the process in which all hydroxy groups are converted to methoxy groups.

The reaction begins with interaction of the glycoside with a methoxy group bound to a leaving group (Tf, or trilate, is a good leaving group). Exhaustive methylation of the monosaccharide occurs. In the presence of acid, the alcohol group is lost and a polymethylated reducing sugar results.

C. ACETYLATION: ESTER FORMATION

Treatment of a sugar with an excess of acetic anhydride and base (such as pyridine) results in polyacetylated sugar. The reaction, carried out at low temperatures, is stereospecific (i.e., α gives α-esters). These groups can be used for the **stereoselective formation of glycosides** (Figure 7-6).

The reaction starts with the formation of a polyacetylated sugar by reacting a simple sugar with acetic anhydride and base. The acetic anhydride is the source of the acetyl groups that add on to the sugar -OH groups. Treatment of the polyacetylated sugar with HBr results in the displacement of the acetyl group on the anomeric carbon. The anomeric carbon is attacked because it is electrophilic. Bromine is a good leaving group and, under S_N1 conditions, Br leaves to give the carbocation. This carbocation is stabilized by the lone pair electrons that the ring oxygen donates through resonance. The carbocation is further stabilized by the lone pair electrons of the neighboring acetyl group oxygen. This interaction protects the bottom face of the carbocation from attack, resulting in exclusive β-anomer formation because the alcohol can attack only from the top face of the carbocation.

III. Polysaccharides

Also known as glycans, polysaccharides consist of monosaccharides connected by glycosidic linkages of many varieties. They can either be α or β and they can occur between **var-**

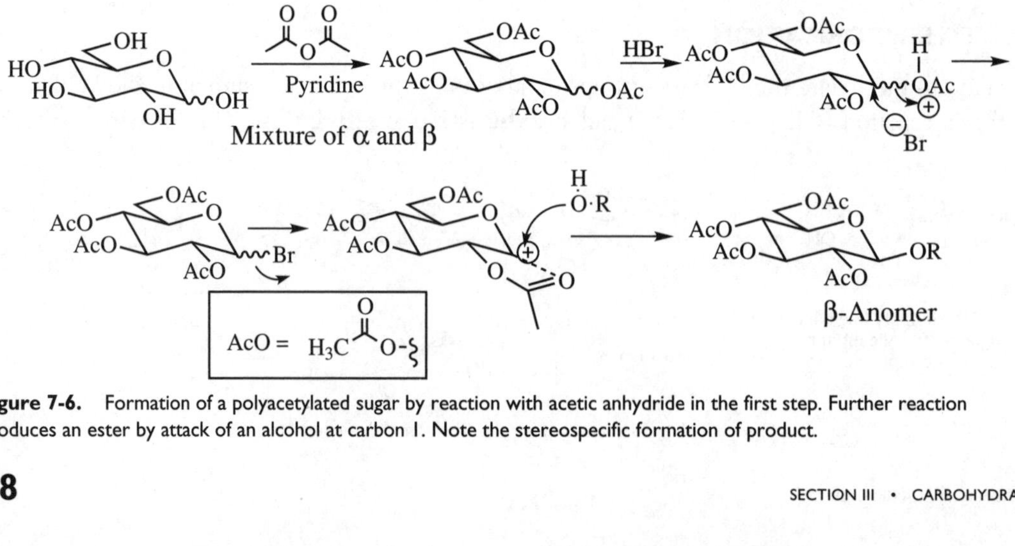

Figure 7-6. Formation of a polyacetylated sugar by reaction with acetic anhydride in the first step. Further reaction produces an ester by attack of an alcohol at carbon 1. Note the stereospecific formation of product.

68

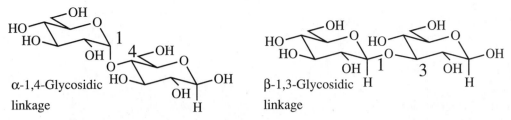

Figure 7-7. Glycosidic linkages. Do not memorize the structure of polysaccharides, but understand the concepts involved.

ious carbon hydroxyl groups. For example, the hydroxyl group of carbon 1 of one sugar can bind to the hydroxyl group of carbons 3, 4, 6, and so on of another sugar.

In Figure 7-7, an α-1,4 bond connects the carbon 1 α-hydroxy group of one sugar unit to the carbon 4 hydroxy group of another. Also shown is a β-1,3 glycosidic linkage, which connects the carbon 1 β-hydroxy group of one sugar unit to the carbon 3 hydroxy group of another sugar.

Starch consists of glucose molecules connected with α-1,4 bonds. These bonds can be hydrolyzed easily to yield monosaccharides. **Glycogen** is also a polysaccharide, but unlike starch, it is also highly branched. Some of the sugar units may contain 1,4 and 1,6 linkages, resulting in a branch point. **Cellulose** (fiber) is composed of β-linkages that cannot be broken down by human enzymes; it consists of D-glucose with β-linkages. This conformation is energetically favorable because the sugars are in chairs with the bulky groups in equatorial positions. **Glucosamines** are linkages of carbon 1 with an amine. Examples include nucleotides that link ribose and deoxyribose to bases (RNA, DNA).

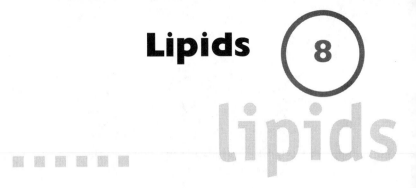

Lipids ⑧

I. Structure

Lipids may be isolated from cells and are soluble in nonpolar organic solvents (chloroform, diethyl ether). Lipids include compounds of many different structures and functions: **fatty acids, triacylglycerols, terpenes, phospholipids, prostaglandins, waxes, and others.**

A variety of lipids are illustrated in Figure 8-1. At the far left of Figure 8-1 is the basic structure of a **fatty acid,** which contains a carboxyl group and a long hydrocarbon chain. Combining three fatty acids with the molecule glycerol yields the **triacylglycerol** (triglyceride). Note that triacylglycerols are esters of fatty acids.

Also shown in Figure 8-1 is the structure of **isoprene,** one of nature's favorite building blocks. The five-carbon–containing isoprene molecule is a diene (i.e., has two double bonds). Isoprene units can be combined in various ways to create **terpenes,** molecules that have carbon skeletons consisting of isoprene units joined head to tail. Terpenes are found in the oils of many plants. The terpene compound in Figure 8-1 is an oil from the ginger plant.

Finally, isoprene units can be combined to create complex ring structures. Steroid compounds are made by a step-by-step combination of isoprene units; the **steroid skeleton** appears at far right in Figure 8-1.

II. Fatty Acids and Triacylglycerols

A. STRUCTURE

Fatty acids consist of a carboxylic acid group attached to a long hydrocarbon chain. If this chain does not contain multiple bonds, it is referred to as **saturated.** If the chain has multiple bonds, the structure is **unsaturated.** Triacylglycerols, esters of fatty acids and glycerol, are the form in which energy is converted for long-term storage in fat cells. Unsaturated fats (vegetable oils) are therefore triacylglycerols with unsaturated fatty acid hydrocarbon chains. The double bonds are all *cis* and generally are unconjugated. Saturated or animal fat has been implicated in the development of heart disease.

Fats are used for storage because they release more than twice the energy per gram compared with carbohydrates. This difference is attributed to the larger number of C–H bonds in fats per molecule (and to less hydration per gram). At room temperature, saturated fats generally are solids (e.g., butter), whereas unsaturated fats typically are liquids (e.g., corn oil).

B. SAPONIFICATION

Saponification involves the hydrolysis of the ester linkages of glycerides (Figure 8-2), which results in the release of the fatty acids as salts and glycerol. **The long-chain fatty acid salts are soaps.** This reaction is part of the process by which many soaps are made.

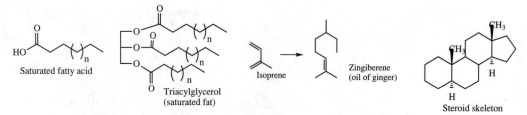

Figure 8-1. Examples of lipids: fatty acid, triacylglycerol, isoprene, terpene, and steroid skeleton.

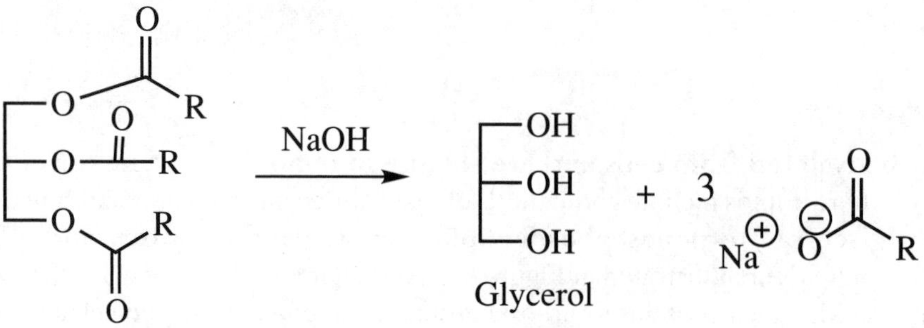

Figure 8-2. Saponification of triglyceride to glycerol and fatty acid salts (soaps).

The long-chain fatty acids contain a hydrophilic (likes water) polar head group with a long hydrophobic (shies away from water) tail. At specific concentrations in aqueous solutions, these fatty acids lead to the formation of micelles.

Micelles are spherical structures made up of hundreds of fatty acid soap molecules (Figure 8-3). Micelles are arranged with the polar group of the fatty acid soap molecules on the outside and the hydrophobic chains embedded on the inside, away from the water. The structure of the micelle explains how soap works. Micelles trap dirt and grease (which are hydrophobic) in the center of the micelle. The micelles are soluble in water, because the surface of the micelle contains the polar carboxylate group. Thus, micelles can wash away with water and carry dirt and grease with them.

C. PHOSPHOLIPIDS

Phospholipids are structures in which one of the fatty acid chains in a triacylglycerol is replaced by phosphoric acid. Therefore, phospholipids look much like triglycerides. Phosphoric acid (PO_4H_3) is a strong acid that has three free -OH groups and a free oxygen atom that carries a negative charge. Phosphoric acid has the capacity to form a total of three ester bonds— an ester bond at each of its hydroxyl group sites. When an ester linkage occurs at two of its -OH

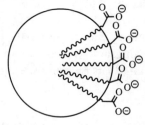

Figure 8-3. A micelle. Note the polar carboxylate groups on the outside and the nonpolar hydrocarbon chains on the inside of the structure.

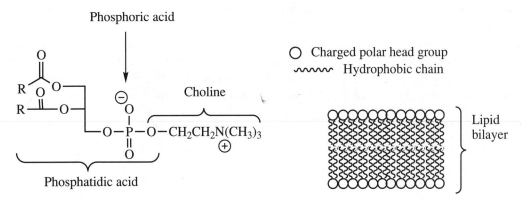

Figure 8-4. Phosphatidic acid and choline combine to form phosphatidyl choline, an important phospholipid (left). The lipid bilayer (right) is formed by phospholipids.

group sites, **phosphodiesters** form. (See Chapter 11 to review how phosphoric acid can form ester linkages.)

When phosphoric acid is linked to the glyceride by forming an ester bond at one of its -OH groups, the resulting molecule is **phosphatidic acid** (Figure 8-4). When phosphatidic acid is linked to an alcohol molecule, such as **choline,** by an ester linkage, the result is a **phosphodiester linkage** and a molecule known as **phosphatidyl choline.**

The phospholipid found in cell membranes is phosphatidyl choline. The structure of this molecule (see Figure 8-4) is important because the choline/phosphate part is polar and the fatty acid chains are nonpolar. Thus, these compounds have two hydrophobic chains and a polar head group with a positive (amine) and a negative charge (phosphorus). This **"amphoteric"** nature allows for the formation of micelles and the formation of a **lipid bilayer** (see Figure 8-4), which is a critical part of cell membranes.

III. Steroids

These compounds include a wide range of molecules that are essential to living organisms: **cholesterol, the sex hormones (the estrogens, testosterone, progesterone), adrenal cortex hormones, and all of their metabolites** (Figure 8-5).

Steroids are based on a skeleton with four basic rings (labeled A–D in Figure 8-5). For most steroids, the junction between rings B,C and C,D are *trans* to each other. They are termed 5α or β because the type of ring junction is determined by the stereochemistry of the carbon 5 position. Do not memorize the steroid structures; instead, understand that **the diversity of steroid com-**

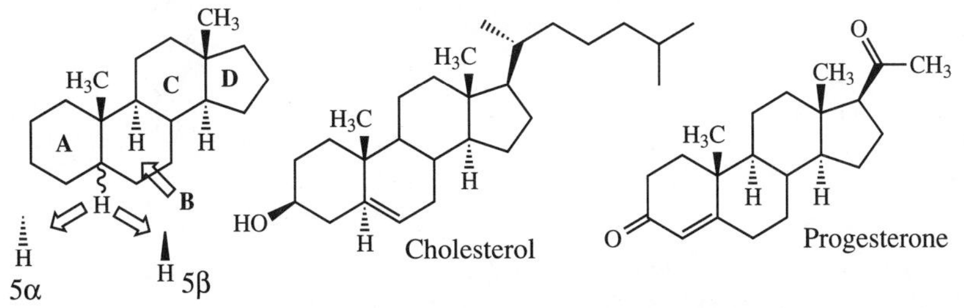

Figure 8-5. The steroid skeleton (left), and two important steroids, cholesterol and progesterone (right).

pounds is based on a variety of possible ring junctions and different substituents on the basic steroid skeleton.

Steroid hormones work by passing through cell membranes to bind to intracellular receptors of the cell. Binding initiates a cascade of chemical processes, usually within the nucleus of the cell. The reactions of steroids are as varied as the different types of structures. The functional groups attached to the steroid skeleton can undergo all of the reactions reviewed in the previous chapters.

Amines  9

■ ■ ■ ■ ■ ■

I. Structure and Physical Characteristics

The most important characteristic of amines is the lone pair of electrons. The nitrogen of amines is sp^3 hybridized and the lone pair occupies one of the sp^3 hybridized orbitals. The result is a **trigonal pyramidal** structure with the substituents at the base of the pyramid (Figure 9-1).

With four different substituents (one being the lone pair electrons), it would seem that two separate enantiomers of amines could occur. A rapid **interconversion** takes place, however, and the energy required for this conversion is relatively low. Interconversion occurs readily at low temperatures, which means that the enantiomers cannot be resolved. The process of interconversion, **through an sp^2 hybridized transition state,** is depicted in Figure 9-1.

The melting and boiling points of amines are high because of hydrogen bonding. The hydrogen bonding patterns are similar to the hydrogen bonding pattern seen in water and alcohols. The effects of branching and chain length are the same as for alkanes (see Chapter 2).

II. Basicity

A. OVERVIEW

Bases are proton acceptors and electron donors. **Amines are basic because they can donate the lone pair electrons to protons.** The basicity constant defines base strength (Figure 9-2). **A larger base constant indicates that more of the species exist in the protonated form, resulting in a better base;** pK$_b$ = $-$log K$_b$, so the smaller the pK$_b$ value, the stronger the base.

Aryl amines tend to be weaker bases. In the nonprotonated state, the molecule is stabilized by resonance with the benzene ring. When the lone pair is donated to a proton, however, it can no longer delocalize through the benzene ring and the stability of delocalization is lost (Figure 9-3).

The same type of effect decreases the basicity of amide nitrogens. The nitrogens are involved in delocalization through the carbonyl. If protonation occurred, the stabilizing resonance interaction would not occur (Figure 9-4). Thus, the amide nitrogen is not basic. In actuality, when reacted with acids, the carbonyl oxygen is protonated because of the higher electron density about the oxygen.

Also because of amide nitrogen's basicity and the lone pair, **nitrogen atoms are able to stabilize adjacent carbocations,** making them easier to form. The lone pair occupies an sp^3 orbital in an sp^3 hybridized nitrogen, which lines up with the empty p orbital of the adjacent carbocation and donates electrons to the C-N bond to help **spread out the charge.**

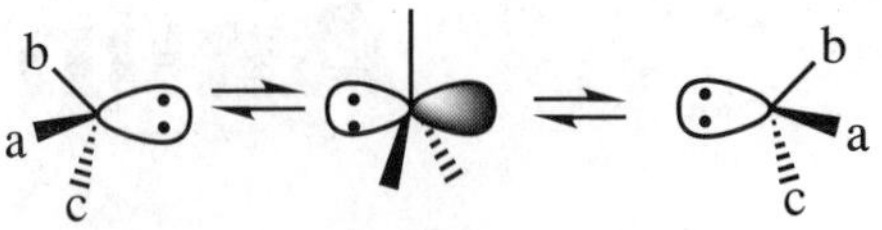

Figure 9-1. The sp^3 hybridized structure of amines. The lone pair of electrons occupying one of the orbitals results in a trigonal pyramidal structure. Note also the interconversion through a transition state (see text for explanation).

$$RNH_2 + HOH \rightleftharpoons \overset{\oplus}{R}NH_3 + \overset{\ominus}{O}H \qquad K_b = \frac{[\overset{\oplus}{R}NH_3][\overset{\ominus}{O}H]}{[RNH_2]}$$

Figure 9-2. Relationship for K_b.

B. GOVERNING FACTORS

The more highly substituted the amine with electron releasing-alkyl groups, the more basic the amine (Figure 9-5). The electron-releasing groups can stabilize the positive charge of the substituted ammonium ion. Thus, trimethylamine is a stronger base than ammonia. It is generally true that **aromatic amines are weaker bases than ammonia, given the resonance stability of the aromatic compounds.**

Hybridization also effects basicity. Consider piperidine and pyridine (Figure 9-6). Both structures possess lone pair electrons on the nitrogen, although the electrons in piperidine are in an sp^3 orbital, whereas those in pyridine occupy an sp^2 orbital. The increased s character of the sp^2 orbital places the electrons closer to the nucleus and therefore they are held more tightly by the nitrogen. Thus, piperidine is a stronger base because its electrons are more "available" for proton acceptance.

When reacted with acids, followed by repeated alkylation, amines form quaternary ammonium salts, which have the formula $R_4N^+X^-$ (X is a halogen ion). For example, if the amine RNH_2 reacts with acid (RX), R_2NH is produced. Continued alkylation with acid (RX) produces R_3N, resulting finally in the quaternary ammonium salt $R_4N^+X^-$.

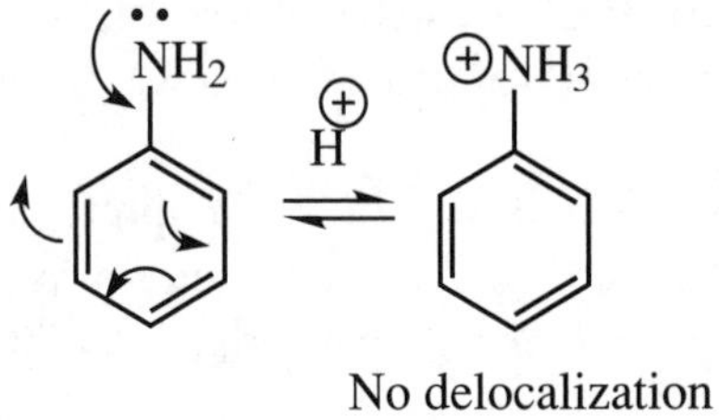

No delocalization

Figure 9-3. When nonprotonated, the aryl amine is stabilized by resonance. When protonated, the lone pair is donated to the proton and delocalization is lost.

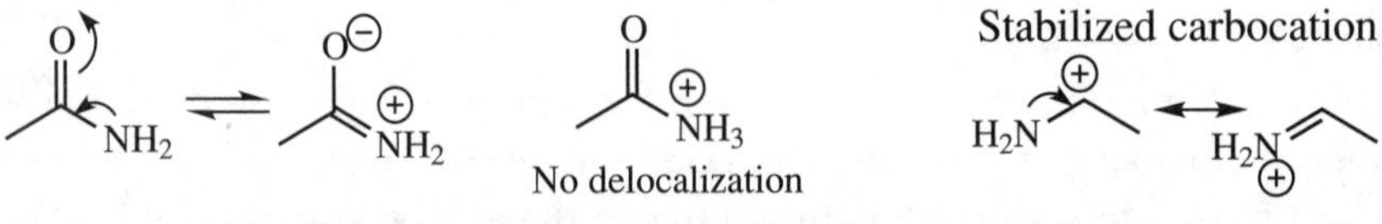

No delocalization   Stabilized carbocation

Figure 9-4. Resonance forms associated with the amide bond (left). Protonation of the amine group and loss of delocalization (center). Stabilization of carbocations by nitrogen (right).

$$(CH_3)_3N > (CH_3)_2NH > CH_3NH_2 > NH_3$$

Figure 9-5. Decreasing order of base strength based on substitution of electron-releasing groups.

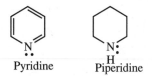

Pyridine   Piperidine

Figure 9-6. Pyridine and piperidine. Piperidine is the stronger base because its electrons are more available for proton "acceptance."

III. Reactions

A. NUCLEOPHILIC ATTACK BY AMINES

Recall that in the attack of electrophiles by amines, it is the **lone pair electrons that allow the amine to act as a nucleophile. Ammonia (with no substituents) and primary amines make the best nucleophiles because of the lack of steric hindrance.** As the level of substitution increases, the lone pair electrons are less able to reach the electrophile to form a bond, and the amine acts more as a base than a nucleophile. As discussed previously (see Chapter 6), amines can undergo an S_N2-type attack of electrophiles and an attack of carbonyls to form imines and amides (Figure 9-7).

B. PREPARATION OF AMINES: REDUCTION

As detailed in Chapter 6, the **reduction of nitriles and amides by reducing agents such as lithium aluminum hydride ($LiAlH_4$) results in the corresponding amines. Imines can also be reduced to amines.** This process is called **reductive amination.** Alkyl azides ($R-N_3$) can also be reduced, using $LiAlH_4$, to yield the amine (RNH_2).

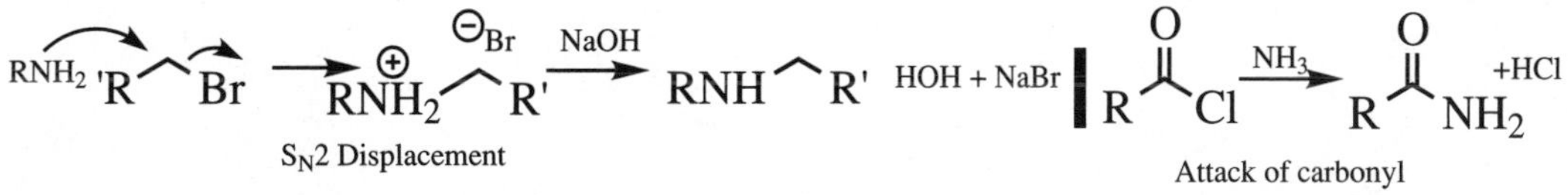

Figure 9-7. Nucleophilic attack by amines. S_N2-type attack with good leaving group (left); attack of a carbonyl to form an amide (right).

Amino Acids and Peptides (10)

I. General Structure of Amino Acids

Of the different types of biopolymers (polysaccharides, nucleic acids, proteins), proteins have the most diverse functions. They can act as **hormones,** which bring about a specific physiologic response; enzymes that catalyze chemical reactions; and **structural elements** within our body (bones, skin, muscle); and they can be involved in transport, regulation, and a host of other processes. **Proteins are polymers of amino acid units.** Proteins can have hundreds of amino acids linked linearly by amide bonds to form complex three-dimensional structures. **Peptides,** which are smaller polymers of amino acids, often act as hormones. The specific three-dimensional structure of peptides allows for recognition of the hormone by a **receptor.** Thus, the structural and physical characteristics of the proteins and peptides are determined by the structural and physical characteristics of the basic amino acid building blocks.

Twenty naturally occurring *a*-amino acids are used by cells to synthesize peptides and proteins (Figure 10-1). **The primary structure of a protein is the specific linkage sequence of these amino acids.** This primary structure must be intact for the protein to function normally. A specific primary structure allows interaction of the individual amino acids in the protein such that the protein may fold in a specific manner to give rise to secondary and tertiary structure.

The amino acids differ from one another on the basis of the R-group side chain structure, which is attached to the basic amino acid skeleton. In Figure 10-1, note that each of the R-groups attaches to the amino acid skeleton to generate the 20 different naturally occurring L-amino acids.

The *a*-amino acids are named because of the NH₂ group attached to the *a*-carbon of the carboxylic acid. Like carbohydrates, they are also given the D or L designation (L-amino acids have the amino group on the left of a Fischer projection, D-amino acids have the amino group on the right) [Figure 10-2]. Also, as with sugars, whether an amino acid is characterized as D or L says nothing about the direction it will rotate plane-polarized light (*d* or *l*).

In most cases, D-amino acids are *R* (absolute configuration) and L-amino acids are *S*. Remember that this designation depends on the side chain priority. Amino acids are similar to sugars in that D- and L-forms of a particular amino acid are enantiomers. The exceptions to this rule include the amino acids threonine and isoleucine, which contain a chiral center on the side chain, making the D- and L-forms diastereomers. Glycine has no chiral center at the α-carbon so there are no D- or L-forms. In most naturally occurring proteins, L-amino acids predominate.

II. Classification of Amino Acids

The amino acids are classified into five different groups according to the reactivity and characteristics of the side chains (see Figure 10-1).

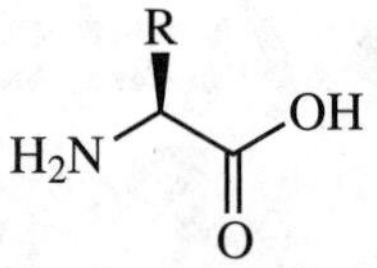

General Structure for L-amino acids

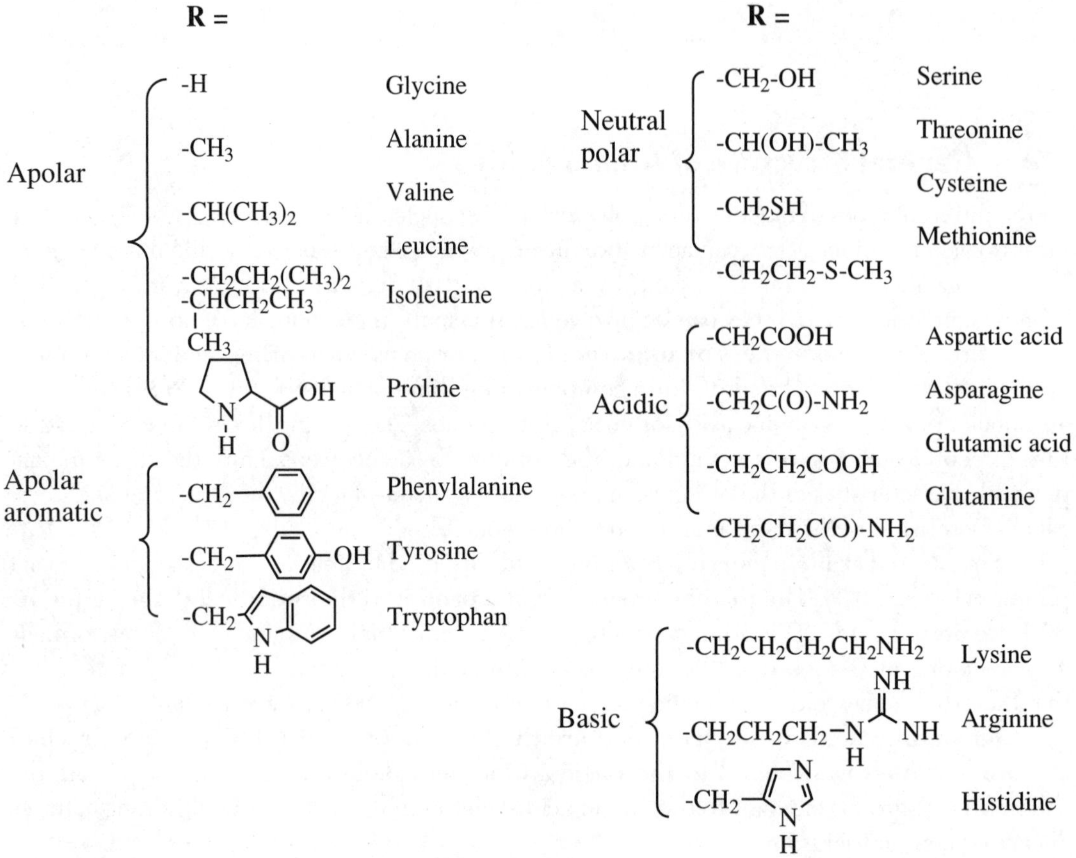

R =   **R =**

Apolar

- -H — Glycine
- $-CH_3$ — Alanine
- $-CH(CH_3)_2$ — Valine
- Leucine
- $-CH_2CH_2(CH_3)_2$
- $-CHCH_2CH_3$ | CH_3 — Isoleucine
- Proline ($-OH$)

Apolar aromatic

- $-CH_2-$⬡ — Phenylalanine
- $-CH_2-$⬡$-OH$ — Tyrosine
- $-CH_2-$(indole) — Tryptophan

Neutral polar

- $-CH_2-OH$ — Serine
- $-CH(OH)-CH_3$ — Threonine
- $-CH_2SH$ — Cysteine
- $-CH_2CH_2-S-CH_3$ — Methionine

Acidic

- $-CH_2COOH$ — Aspartic acid
- $-CH_2C(O)-NH_2$ — Asparagine
- $-CH_2CH_2COOH$ — Glutamic acid
- $-CH_2CH_2C(O)-NH_2$ — Glutamine

Basic

- $-CH_2CH_2CH_2CH_2NH_2$ — Lysine
- $-CH_2CH_2CH_2-N-\overset{NH}{\overset{\|}{C}}-NH$ H — Arginine
- $-CH_2-$(imidazole) — Histidine

Figure 10-1. The 20 α-amino acids classified by R-group. Note the classification groups and the side chain types.

A.   APOLAR

These amino acids, with the exception of glycine, contain hydrophobic side chains and are highly inert. They have a profound effect on protein conformation because they tend to aggregate (owing to Van der Waals interactions) and shy away from water. Thus, they are commonly found on the interior of proteins. Proline has a unique cyclic **imino** group and is found in regions of the amino acid chain where a turn occurs.

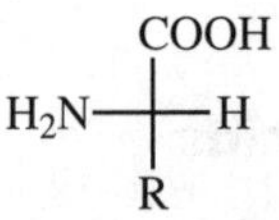

COOH
H₂N——H
R

Figure 10-2. Fischer projection of the amino acid skeleton. Note the amino group on the left, and the L-amino acid designation.

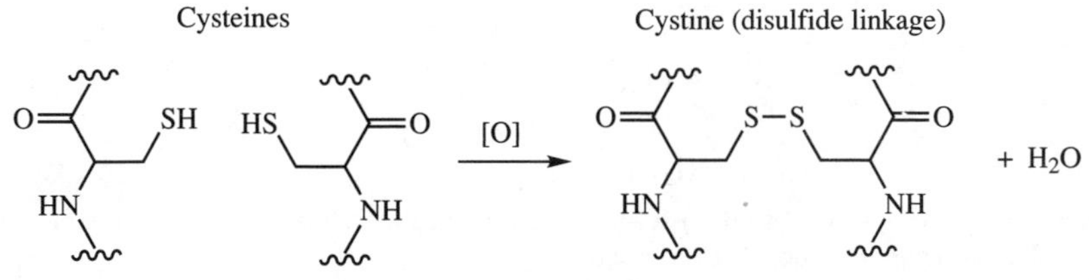

Figure 10-3. Disulfide linkage formation.

B. APOLAR AROMATIC

These side chains are also hydrophobic and relatively inert, tending to aggregate because of Van der Waals interactions. The tyrosine side chain contains a slightly acidic phenolic hydroxy group that is often involved in **hydrogen bonding.** Tryptophan contains the highly hydrophobic indole ring in which the lone pair electrons of the nitrogen contribute to the aromatic system to give 10 π electrons [remember the formula $(4n+2)\pi$ electrons].

C. NEUTRAL POLAR

These side chains are uncharged at physiologic pH, but they can also be somewhat reactive. The hydroxy group of serine is often used as **a nucleophile** in enzyme active sites. The hydroxyl group of serine can participate in hydrogen bonds. The thiol group of cysteine is also a nucleophile, but it plays a more important role in **disulfide linkages.** On treatment with a mild oxidant, disulfide linkages (cystine) can form from two cysteines with side chains in proximity. The sulfur is a large polarizable (fluffy) atom, and thus it is able to deprotonate more readily than oxygen (thiol is a better acid than hydroxy), allowing for the easy oxidative coupling of the two atoms. Disulfide bonds (Figure 10-3) can form in proteins between cysteine molecules and often help stabilize tertiary structure (see subsequent discussion). Threonine and methionine also are polar but they are far less reactive.

D. ACIDIC SIDE CHAINS

The side chains of glutamic acid and aspartic acid are usually deprotonated at physiologic pH and thus carry a negative charge. **They often play a large role in the active site of enzymes.** Remember that these amino acids are acidic. They may act as proton donators (protonated form) or as proton acceptors (deprotonated form). Glutamine and asparagine are not truly acidic, but are placed in this group because they are derivatives of glutamic and aspartic acids; they tend to be good hydrogen bond donors and acceptors.

E. BASIC SIDE CHAINS

The side chains of these amino acids are also frequently involved in the active site of enzymes. The side chain of arginine **(guanidine group)** is basic. At physiologic pH, this group is always protonated (Figure 10-4). The aromatic side chain of histidine has an **imidazole ring** (see Figure 10-4), which can act as an acid or base and is a good nucleophile. Note that both side chain functions shown in Figure 10-4 can be stabilized by resonance on protonation by an acid.

III. Amino Acids as Dipolar Ions

Amino acids in the dry solid state exist as **zwitterions or dipolar ions,** which means the amino group is protonated and the carboxyl group is deprotonated (NH_3^+, COO^- on the same

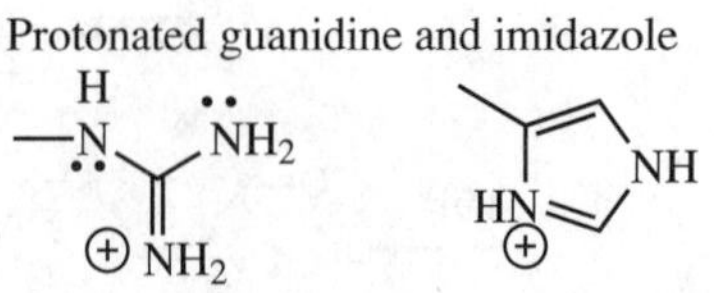

Protonated guanidine and imidazole

Figure 10-4. Side chain group guanidine (arginine) at left; side chain group imidazole (histidine) at right. Note the possible resonance for each of these important basic side chain groups.

molecule). In solution, amino acids exist in equilibrium with protonated and deprotonated states. It is this characteristic of amino acids and their side chains that gives them their unique functional characteristics in peptides and proteins.

Whether the amino group, carboxyl group, and side chain group are protonated depends on the pH of the solution. **The pK_a of the carboxyl group is about 2.3. The pK_a of the amino group is generally about 9.7.** Therefore, at a pH of 2.3, about one half of the carboxyl groups of an amino acid are protonated. You would expect all of the amino groups to be fully protonated at this low pH. At pH 9.7, one half of the amino groups are deprotonated. You would expect all of the carboxyl groups to be deprotonated at this relatively high pH.

The amino acid alanine (Figure 10-5) can exist in a cationic, dipolar, or anionic state, depending on the pH of the solution. In acidic solution (pH < 2), alanine is in a cationic state because its carboxyl group is protonated and its amino group is protonated and carries a positive charge. As the pH of the solution containing alanine increases with the addition of base, the carboxyl group begins to deprotonate. By pH 2.3, one half of the carboxyl groups are deprotonated, resulting in the dipolar ion (see Figure 10-5, center). As the pH is further raised by adding more base, the amino group deprotonates next because of its higher pK_a. The deprotonation of the amino group results in alanine becoming an anion.

A. ISOELECTRIC POINT

The isoelectric point is the pH at which an amino acid does not migrate in an electric field. It typically occurs where **the concentration of the dipolar ion is greatest.** In Figure 10-5, the isoelectric point is the average of the two pK_a values, or 6.0. Remember that according to the Henderson-Hasselbach equation (see Chapter 1), when the pH of a solution is equal to the pK_a of an acidic proton, half of the molecules are protonated and half are deprotonated.

B. TITRATION

By placing an amino acid solution in aqueous acid solution (pH 0), adding base, and making a plot of pH versus equivalents of base added, you obtain a graph similar to that for lysine in Figure 10-6. Lysine has a carboxyl and amino group like all amino acids, but also has a basic side chain group containing an $-NH_2$ group. This $-NH_2$ group can be protonated and deprotonated.

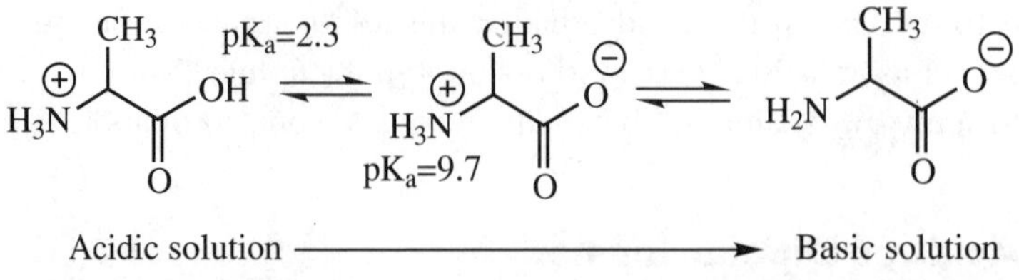

Figure 10-5. Addition of base to a fully protonated amino acid (alanine). Note that in acidic solutions, the amino acid is fully protonated. As the pH is raised by adding base, deprotonation of the carboxy group occurs first, producing the dipolar ion. Further deprotonation of the amino group occurs at higher pH.

82

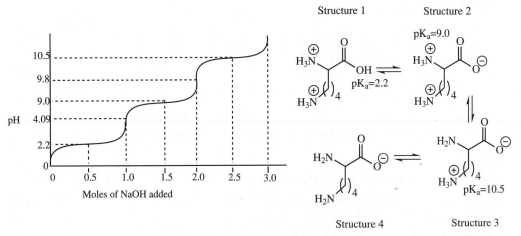

Figure 10-6. Titration curve for the amino acid lysine. The fully protonated form deprotonates first at its carboxyl group, followed by deprotonation of its amino group and side chain basic group. Structures are shown for each step of deprotonation.

The first structure in Figure 10-6 shows lysine fully protonated. At pH 0, lysine exists as a dication because both the amino group and basic side chain NH_3 group are protonated. All the molecules look like structure 1. As base is added, the carboxyl group becomes deprotonated. After one full equivalent of base is added, (1 mole of NaOH), all of the carboxyl groups are deprotonated and all lysine molecules look like structure 2. Note that a rapid rise in pH has occurred. As more base is added, the α-amino group is deprotonated, resulting in the dipolar ion. By the time a full second equivalent of base has been added, the lysine molecules look like structure 3. As even more base is added, the side chain amine group is deprotonated, leading to the anionic form of lysine in structure 4. Lysine would be expected to be completely in a form similar to structure 4 when a total of three equivalents (3 moles of NaOH) of base are added.

The pK_a of the carboxyl group is found at the center of the first plateau of the graph in Figure 10-6, at about pH 2.3. The pK_a of the α-amino group is at the center of the second plateau of the graph, at about pH 9.0. By looking at the pH of the third plateau, you can predict the pK_a of the side chain amine group to be about 10.5. Remember that at each pK_a, about one half of the molecules are deprotonated for the group under consideration. Note that the pK_a for the α-amino group of lysine (9.0) is lower than most other amino acids (9.7).

IV. Formation of Peptide Amide Bonds

Different ways to make amide bonds have been discussed previously. In general, an amine is used to attack a carboxylic acid derivative that has a good leaving group, and an amide is formed. To synthesize peptides, it is often easiest to couple amino acid fragments. The carboxylic acid of one residue is activated for attack (converted to an activated carboxylic acid derivative) and the amine of the other residue is used to attack it. To ensure that the right connections take place, **pro-**

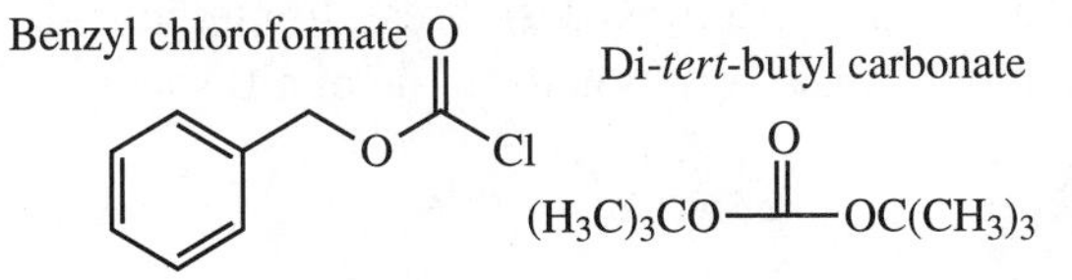

Figure 10-7. Protecting groups used to help form desired peptide bonds. Do not memorize these structures, but understand why they may be useful.

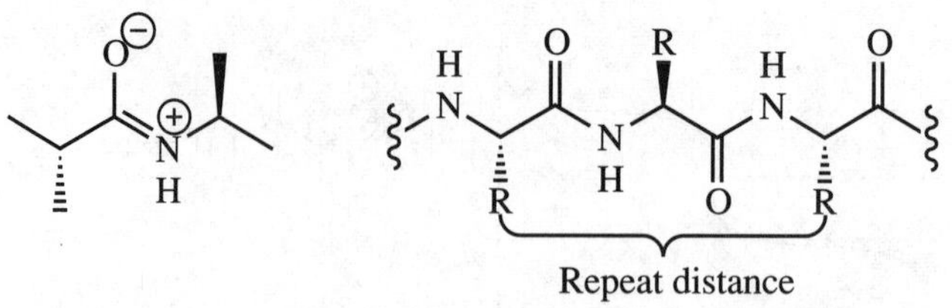

trans Amide

Figure 10-8. The *trans* form of the amide bond (left) and the repeat distance (right).

tecting groups are linked to reactive groups that are not meant to participate in the formation of amide bonds. Without this step, products other than the desired connection are possible.

Several types of protecting groups are available, including *tert*-butyl-based groups and benzyl-based groups (Figure 10-7).

V. Secondary and Tertiary Structure

A. SECONDARY STRUCTURE

Secondary structure is dictated by the *trans* nature of the amide bond and by hydrogen bonding along the amide-bonded main chain of the peptide or protein. Two examples of secondary structure are the β-pleated sheet and α-helix. Recall that through resonance, the C–N bond of an amide assumes an almost double-bond–like character. Rotation about this C–N bond is more restricted than that about a normal C–N single bond. As in alkenes, **the *trans* structure is more favored than the *cis* structure because of the higher eclipsing strain of the *cis* structure** (Figure 10-8). When the amino acid chain is spread out, the *trans* nature of the amide bonds make the side chains alternate in terms of from what side of the chain they protrude. The **repeat distance** is the distance between two residues appearing on the same side of the chain; in linear form, it is approximately 7.2 Å.

1. TURNS

The peptide chains are not in straight lines. In many cases, the chains have up to 180° changes in direction, which are stabilized by hydrogen bonds between the proton of N-H and the oxygen of the carbonyl. The inclusion of **proline** in a peptide chain allows for *cis* amide bonds to form in the residue before it. Thus, proline is often found in regions where the turns occur. **Glycine** is also frequently found in turn structures because it does not have a side chain to sterically hinder a turn structure from forming around it.

2. β-SHEETS

Antiparallel β-sheets often occur between two or more linear chains that are in proximity. The term **antiparallel** is applied because the adjacent chains of a β-sheet run in opposite directions. **These chains are held together by hydrogen bonding** between hydrogen on nitrogen and the carbonyl oxygen. These sheets are also pleated (like a room divider) to avoid steric interactions between the side chain groups (Figure 10-9). Small to medium-sized R-groups can be accommodated inside the sheet. Silk is a common example of a β-sheet structure.

3. α-HELIX

This structure shown in Figure 10-9 is a **right-handed helix stabilized by hydrogen bonding of the amide hydrogens and carbonyl oxygens (dotted lines).** Note the 3.6 amino acids per

84

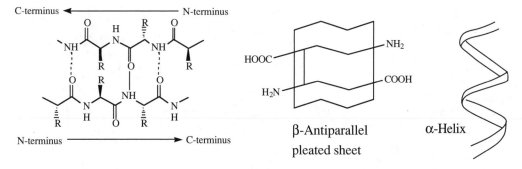

Figure 10-9. Secondary structure: the β-pleated sheet and α-helix. Dotted lines indicate hydrogen bonding.

turn and a rise of 5.4 Å per turn. Each amide has a hydrogen bond to another amide three residues above and below. The repeat distance is much shorter (1.5 Å) than in the β-sheet because the distance along the axis of the helix is used for measurement. In this structure, the side chain groups extend out of the helix, so steric interaction between the groups is minimal.

B. TERTIARY STRUCTURE

Tertiary structure is the arrangement of secondary structure into a specific three-dimensional form caused by the interactions between side chains that protrude from the secondary structure. These folding processes do not occur randomly. Often, if certain residues are mutated (replaced by others or left out), the correct structure does not form and the function of the protein is destroyed.

The arrangement of these globular proteins is dictated by several factors:

1. Residues with nonpolar hydrophobic side chains are often found in the interior of the protein, away from water.
2. Polar charged side chains are often found near the surface of globular proteins and are in contact with the aqueous environment. When in the interior, they may form salt bridges (protonated amine forming salt with deprotonated carboxyl) to stabilize tertiary structure.
3. Polar uncharged side chains can be found on the inside or near the surface. When in the interior, they often form hydrogen bonds with other functional groups to help stabilize tertiary structure.

Tertiary structure is held together by a variety of interactions of varying strength. These include:

1. **Hydrophobic interactions**
2. **Hydrophilic interactions**
3. **Hydrogen bonds**
4. **Disulfide linkages**
5. **Van der Waals forces**

Individually, some of these interactions are weak. When thousands of these interactions occur together, however, a large protein can be held together.

VI. Enzyme Action

The enzyme **active site** is a catalytic cavity inside the enzyme where reactions, such as hydrolysis of amide bonds (Figure 10-10), can occur.

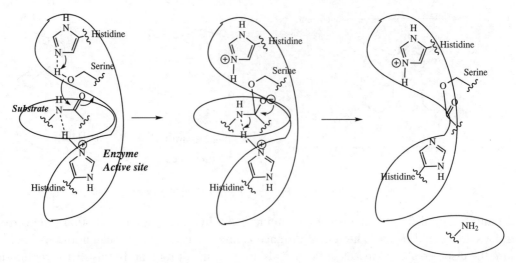

Figure 10-10. Enzyme action.

First, the substrate enters the catalytic cavity and is properly aligned. In the hypothetic active site in Figure 10-10, a neutral histamine (basic imidazole group) partially deprotonates a nearby serine hydroxy side chain. This hydroxy group can attack the correctly positioned carbonyl, and the tetrahedral intermediate is formed. The C–N bond is cleaved as the nitrogen accepts a proton from another nearby protonated histidine (converting the histidine side chain to a neutral basic imidazole). In subsequent steps, a similar mechanism featuring the attack of water on the carbonyl carbon releases the other half of the cleaved amide bond and restores the serine hydroxy side chain. The enzyme is ready to accept a new substrate molecule.

Remember: enzymes are catalysts. They lower the activation energy of a process by placing all of the reactive groups in the correct orientations and helping the reaction along with acidic and basic groups. In some cases, they stabilize (bind to effectively) the transition state of the reaction.

Phosphoric Acid

I. Structure and Physical Characteristics

Phosphoric acid is a triprotic acid that plays a key role in several biologically important molecules. It has three acidic hydrogens with pK_a values of 2.2, 7.2, and 12.2, respectively. On deprotonation, the negative charge can delocalize and is stabilized. These compounds are more acidic than carboxylic acids but less acidic than sulfonic acids. They are anionic in neutral aqueous solutions, as would be expected from their pK_a values.

Monodiesters, diesters, and triesters of phosphoric acid can form when the acid is reacted with an alcohol (Figure 11-1). The mono- and di-linked phosphoric esters are highly acidic because of their remaining hydroxy groups. This acidity gives them stability in aqueous base because the negative charges on the oxygens repel the negatively charged nucleophile (^-OH).

When two or more phosphoric acid molecules link, phosphoric anhydrides form. The structure of a phosphoric anhydride is shown in Figure 11-1.

II. Phosphate in Important Biologic Molecules

For biologically important molecules, ester linkages are formed with hydroxy groups that allow the molecules to carry out their specific functions. For example, DNA and RNA ribose sugars are linked together through **phosphodiester bonds** connecting the C5′ and C3′ hydroxy groups (Figure 11-2). This phosphodiester linkage forms the sugar–phosphate backbone of RNA and DNA.

Phospholipids are also the result of phosphodiester linkages between glycerol and a small alcohol, such as choline (see Chapter 8). The charged regions of phosphatidyl choline (nitrogen and the phosphoric ester), along with the long hydrophobic tail, enable these molecules to act as boundaries in cellular systems.

Phosphoric anhydrides, found in adenosine diphosphate (ADP) and adenosine triphosphate (ATP) [Figure 11-3], play an important role in **energy storage in biologic systems.** Energy from the metabolism of glucose is **"stored"** through the chemical transformation of ADP to ATP through an endothermic reaction. This transformation creates a new **phosphoric anhydride bond.** When the energy is needed, the ATP is hydrolyzed enzymatically to yield ADP and energy that was "stored" in the bond. The hydrolysis of the triphosphate to form the diphosphate is highly exothermic. Enzymes act as catalysts, and are necessary to lower the activation energy so that hydrolysis can occur. This results in the release of the energy stored in this bond.

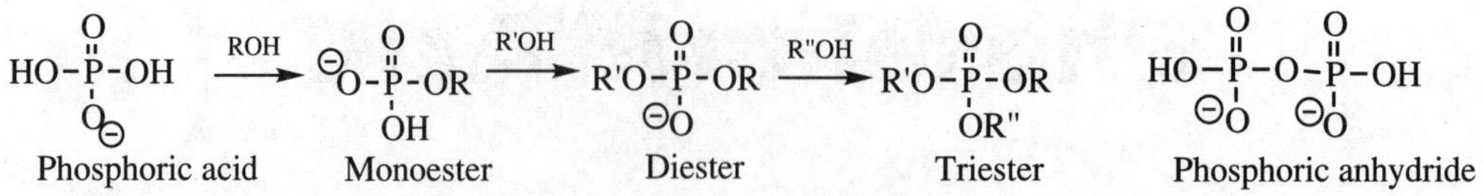

Figure 11-1. Formation of phosphoric acid esters and the structure of phosphoric anhydride.

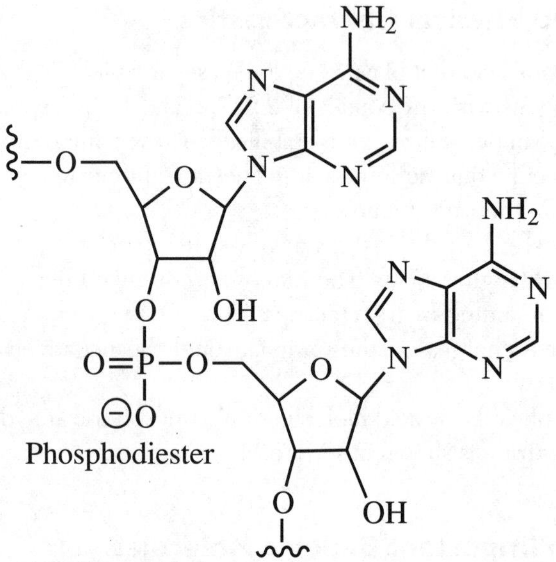

Figure 11-2. Phosphodiester linkage. In the structure of RNA, two ribose sugars are linked together by a phosphate. Note that the ester is formed from the phosphoric acid and two sugar molecules.

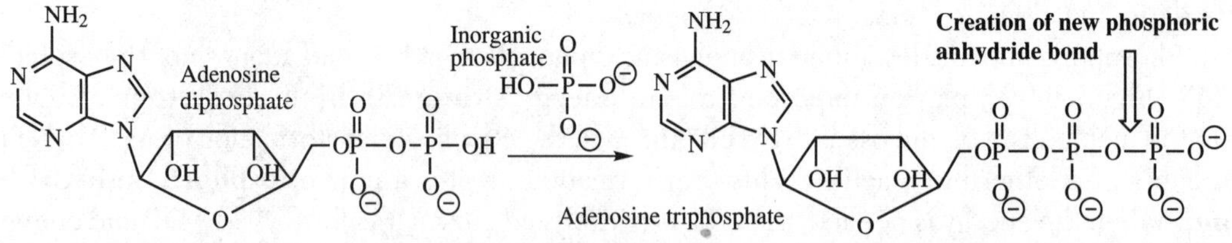

Figure 11-3. Synthesis of adenosine triphosphate (ATP) from adenosine diphosphate (ADP) and inorganic phosphate. A new phosphoric anhydride bond is created that "stores" energy.

Separation, Purification, and Characterization

Clear understanding of the various separation and characterization techniques is essential, not only in terms of their performance, but for the principles behind them.

I. Separation and Purification Techniques

Purification is necessary because reactions often produce undesired side products along with the desired product. Several purification techniques are available.

A. EXTRACTION

Extractions are used to separate compounds that have different solubilities in various solvents. Frequently, the compounds to be extracted are chemically modified to be soluble in aqueous solution or an organic solvent. For example, in a liquid-liquid extraction, two immiscible liquids, such as an aqueous solution and an organic solvent (chloroform, ethyl acetate), are often used to perform the extraction. Separatory funnels with stopcocks are useful in separating the two immiscible liquids.

Key Point: Aqueous acids extract organic bases and aqueous bases extract organic acids.

Figure 12-1 illustrates a coupling reaction of two amino acids linked to a protecting group to form a dipeptide. Note that the protecting group is linked to the carboxyl group of one amino acid and the amino group of the other amino acid. All three compounds, the two amino acids and the dipeptide, are shown. The compounds are dissolved in ethyl acetate, an organic solvent.

By adding an aqueous acid, the amino acid with the exposed amino group becomes a quaternary ammonium salt and is soluble in the aqueous phase (see Figure 12-1, at right). In the ethyl acetate phase, one can find the dipeptide and the amino acid with the exposed carboxyl group. Separation of the layers is easy because ethyl acetate is less dense than water and it is found in the top layer.

By washing the ethyl acetate with an aqueous basic solution, the amino acid with the free carboxyl group is deprotonated (see Figure 12-1, at left) and charged, and becomes soluble in the aqueous layer. It is easy to separate out this aqueous layer with the separatory funnel, leaving only the desired dipeptide in the ethyl acetate, which can be isolated by evaporating the solvent.

B. CRYSTALLIZATION

Recrystallization involves the purification of a solid by dissolving the solid, reducing the volume of the solution with heat, and then cooling the solution. By heating the solution, the solvent evaporates to the point at which the solution is supersaturated. As the solution cools, the solubility rapidly decreases and the compound starts to precipitate.

For recrystallization to be successful, the impurities must at least be soluble in the recrys-

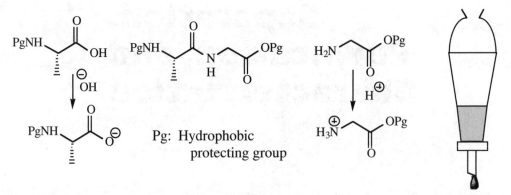

Figure 12-1. Extraction. Two amino acids, linked to a protecting group (Pg), are separated from the dipeptide formed by a coupling reaction. Note that one amino acid is converted to a quaternary ammonium salt (right), whereas the other is deprotonated (left). The amino acids are then easily separated from the dipeptide. A separatory funnel is also shown.

tallization solvent or have a solubility that is greater than that of the desired compound. Otherwise, the impurities crystallize with the desired compound.

C. DISTILLATION

Distillation is often used to purify compounds that have different boiling points. The compounds in their liquid state are heated, and as the boiling point of the compound with the lower boiling point is reached, its vapors are condensed and collected.

Purifying a mixture of several compounds with different boiling points is possible by using a fractional distillation apparatus. In this apparatus, the temperature of the condensing vapor remains relatively constant until most of the material with the lower boiling point is evaporated from the mixture. As the temperature of the condensing vapor continues to rise, the boiling point of the compound with the higher boiling point is reached and the compound vaporizes. These vapors are condensed and collected in a separate flask.

When the separation of the boiling points is large, the distillation can be run under reduced pressure, which reduces the boiling points of the compounds and allows distillation to occur at lower temperatures.

Unfortunately, many liquids have significant vapor pressures at elevated temperatures. Thus, the first vapor that is condensed may contain a portion of the component with the higher boiling point. Repeated distillations may be necessary to increase the purity in these cases.

D. CHROMATOGRAPHY

Chromatography involves the separation of a mixture based on certain differences in the compounds. The exploitable differences include solubility in various solvents and polarity. Chromatography usually involves a stationary and a mobile phase. The **mobile phase** carries the components of a mixture through the stationary phase, and the **stationary phase** hangs onto the components with varying affinities. The different types of chromatography rely on different types of phases, but they are all based on the same concepts.

Chromatography can be used as an analytic tool to monitor reactions or to identify products. It can also be used as a synthetic tool to purify large amounts of material.

1. THIN LAYER CHROMATOGRAPHY (TLC)

TLC is used to **monitor reaction progress and to identify specific components.** This technique is often performed with sheets of glass or plastic coated with the stationary phase (often

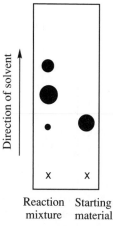

Figure 12-2. Thin layer chromatography.

silicic acid). The liquid mobile phase is a solvent. The mixture to be analyzed is spotted onto the bottom of the plate, and the solvent is allowed to run up the plate by capillary action.

Generally, the stationary phase is polar, and polar compounds stick to the plate more than the nonpolar compounds through dipole–dipole attractive interactions. Polar compounds tend to remain closer to the origin than nonpolar compounds. Nonpolar compounds tend not to adhere to the polar stationary phase and move further up the plate. Thus, **the distance traveled up the plate is a reflection of the polarity of the compound.** An increase in the polarity of the solvent results in less interaction of the compounds with the stationary phase, which allows the compounds in the mobile phase to travel up the plate a greater distance.

The compounds are often visualized by looking at the plate under ultraviolet (UV) light (aromatic compounds absorb UV light) or by dipping the plate in a solution that reacts with the compounds on the plate and stains them.

Figure 12-2 is a typical TLC plate used to monitor a reaction. The reaction mixture is spotted along with the starting material. The solvent is then allowed to run up the plate. The TLC shows that most of the starting material is gone and two products result. These products could then be isolated by column chromatography (see I D 2). In this example, the reaction components are more nonpolar than the starting materials, as shown by the fact that the products travel up the plate farther than the starting material.

2. COLUMN CHROMATOGRAPHY

In this type of chromatography, a column is packed with a stationary phase and the product mixture is loaded on top of the column. The process that follows is essentially the same as TLC, with the exception of the direction of the mobile phase (Figure 12-3). The solvent (mobile phase) runs down the column carrying the mixture while the stationary phase holds on to the compounds with varying affinities. The solvent dripping out the column is collected in fractions, which are analyzed to determine which fraction(s) contains the desired material.

Stationary phases are of several types. **Polar stationary phases** (such as silica gel) bind more tightly to polar compounds, increasing the time polar compounds spend in the column. Not surprisingly, **solvents that are increasingly polar move the compounds more quickly through the column,** because the compounds have less interaction with the polar stationary phase and are retained in the column a shorter amount of time.

Reverse phase columns operate the opposite way: the stationary phase has higher affinity for nonpolar compounds, so the more nonpolar the mobile phase, the more quickly the com-

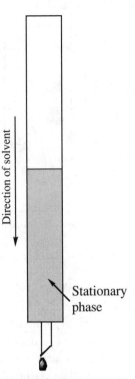

Direction of solvent

Stationary
phase

Figure 12-3. Column chromatography.

pounds go through the column. **In both cases, the retention times can be altered by changing the mobile phase.**

Ion exchange chromatography can be used to separate compounds that are ionic, depending on the pH of the mobile phase. If the desired compound is an amine, the crude solution can be dissolved in an acidic solution to protonate the amine. This solution is then passed through a cation exchange column that binds cations, including the ammonium salt. The other impurities, as long as they are not cations, pass through. The column is then flushed with a salt solution that overwhelms the interactions of the stationary phase with the ammonium salt and releases the desired amine.

The same type of separation can be performed for anions, but with a different stationary phase. To remove an anion, the solution can be passed through an anion exchange column that binds the anions and lets the other compounds flow through.

Size or steric exclusion gel chromatography involves the use of stationary phase beads with little pores that allow small, light molecules to enter while letting larger, heavier molecules pass between the beads. The higher the molecular weight of the molecules, the faster they pass through the column, because the higher–molecular-weight molecules do not have to pass through the small pores in the beads. These columns are often used to purify compounds with very different molecular weights.

3.   GAS CHROMATOGRAPHY

Gas chromatography involves the vaporization of a sample and its movement through a stationary phase. The mobile phase, an inert gas, carries the mixture through a column containing the stationary phase.

The compounds can be detected by a number of different devices after they pass through the column. The results are given as a chromatogram (Figure 12-4), which plots time versus absorbance. This information provides the time the sample came out of the column and the relative concentration (relative peak areas). The relative concentration is proportional to absorbance.

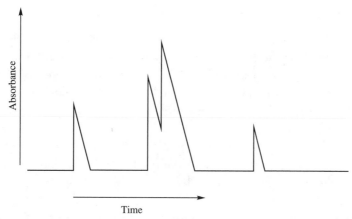

Figure 12-4. Chromatogram produced by gas chromatography or high-pressure liquid chromatography (HPLC) of a compound.

4. HIGH-PERFORMANCE LIQUID CHROMATOGRAPHY (HPLC)

In HPLC, a mobile phase (solvent) runs through a metal-cased column that contains the stationary phase. This procedure is performed at high pressures.

A chromatogram similar to Figure 12-4 is produced by the HPLC procedure. The y-axis of the HPLC chromatogram is often absorbance of UV light, which is proportional to concentration. The area underneath the peak indicates the relative amount of material. The retention time, or the time it takes the compound to go through the column, depends on the flow rate of the mobile phase. In this way, HPLC differs from liquid chromatography, in which the retention time is based more on the composition of the solvent. In the chromatogram in Figure 12-4, the separation is not complete because of the mixing of the two middle peaks. Adjusting the conditions so that retention times are lengthened would result in a better separation.

Key Points: To a point, resolution or distance between the components of a mixture through a column increases as the retention time (time the compounds spend in the column) increases. Longer interaction of the compounds with the stationary phase enhances the separation of the compounds.

Retention time can be increased by:

- **decreasing the speed of the mobile phase**
- **changing the composition of the mobile phase**
- **lengthening the column**

In the laboratory, chromatography can be used to determine the content of a mixture and to monitor the progress of a reaction. For analytic studies, gas chromatography, HPLC, and TLC are most often used because they are **highly sensitive** and require only **a small amount of starting material** to obtain a result. Column chromatography is often used to separate and purify larger amounts of material.

II. Characterization: Infrared (IR) and Nuclear Magnetic Resonance (NMR) Spectroscopy

A. IR SPECTROSCOPY

This form of spectroscopy is commonly used to determine what functional groups are present in a sample. Compared to NMR, however, IR spectroscopy does not provide as much structural information.

1.   THEORY

In previous chapters, bonds have been described as clouds of electrons or molecular orbitals. **For IR spectroscopy, think of the bonds as springs with natural oscillation frequencies.** All bonds have characteristic frequencies with which they stretch and bend. When the frequency of the infrared electromagnetic energy passed through a molecule is equal to the frequency of the stretches or bends of a particular bond, that energy is absorbed. It is this absorbance that can be recorded by an IR spectrometer.

Recall the formula $E = hc/\lambda = h\nu$, which implies that higher frequency means higher energy. Thus, those bonds that require more energy to stretch or bend will have higher absorbance frequencies in IR spectroscopy. The units used in IR spectroscopy are absorbances (in cm^{-1}).

"Stretchiness" depends on the strength of the bond and the masses of the atoms involved in the bond. The hardest bonds to stretch are those with one heavy atom and one light atom. Thus, C–H bonds have high stretching frequencies because of a large mass difference between carbon and hydrogen. Furthermore, for C–H bonds, carbons with triple bonds ($\equiv$C–H) have higher stretching frequencies than carbons with double or single bonds. This information is important, because when you look at an IR spectrum, you can predict what type of C–H bonds are in a molecule. Absorptions greater than $3000 \ cm^{-1}$ indicate multiple-bonded carbons, and absorptions below $3000 \ cm^{-1}$ indicate single-bonded carbons (Table 12-1).

Carbonyl C–O stretches are characteristic and generally occur around 1600 to $1800 \ cm^{-1}$. Amide C=O stretches require the least amount of energy. This low requirement is explained by the partial single-bond character of the C–O bond because of resonance. Single bonds are easier to stretch than multiple bonds.

Hydrogen bonding substituents (hydroxy, amino groups) are also easy to identify because they give a broad absorbance above $3300 \ cm^{-1}$ (O–H or N–H stretch).

2.   TYPICAL IR SPECTRUM

The typical IR spectrum in Figure 12-5 reveals much about the ethyl ester of phenylacetic acid (Ph–CH$_2$CO–OEt). The absorbances above and below $3000 \ cm^{-1}$ indicate the presence of multiple C–H bonds. A strong carbonyl stretch at $1740 \ cm^{-1}$ indicates the presence of an ester. Medium-sized peaks at 1600 and $1500 \ cm^{-1}$ are indicative of an aromatic ring. The C–O stretches are also very strong at 1150 and $1250 \ cm^{-1}$. Clearly, IR can be used as a tool to confirm the structure of a molecule even well after its IR spectrum has been taken and published.

TABLE 12-1. Important Characteristic Absorbances (cm^{-1})

—C—H	2850–2960	—O—H	3400–3640 (very broad)	
=C—H	3020–3100	—N—H	3310–3500 (broad)	
$\equiv$C—H	3300			
—C=C—	1650–1670	C=O	aldehyde	1690–1740
—C$\equiv$C—	2100–2260		ketone	1680–1750
aromatic ring	1600, 1500		esters	1735–1750
			amides	1630–1690
—C$\equiv$N	2210–2260	C—O	ethers, esters	1080–1300

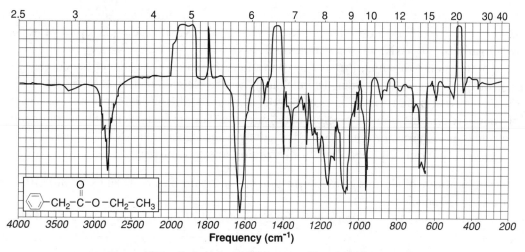

Figure 12-5. Typical infrared (IR) spectrum. Note the multiple bonds.

B. NMR

1. OVERVIEW

The theory of quantum mechanics states that a nucleus with a spin number of I = 1/2 will have two degenerate or energetically equivalent states. In Figure 12-6, **position 1** shows the degenerate or energetically equivalent states. If this spinning nucleus is placed in a magnetic field, these two states will no longer be degenerate, as shown at **position 2.** The lower-energy spin state will have the magnetic moment of the spinning nucleus aligned with the applied magnetic field. The higher-energy spin state will have a magnetic moment aligned against the magnetic field.

Like electrons, the nuclei of certain atoms are considered to spin. The nucleus of the hydrogen atom contains one proton that is considered to spin, and in doing so, is thought of as a tiny bar magnet.

If a sample of protons is placed in a magnetic field, some align with and others align against the applied magnetic field. Alignment with the field is considered more stable. Energy must be absorbed to "flip" the tiny proton magnet over to the less stable alignment, against the field. If these spinning nuclei are hit with electromagnetic radiation at the correct frequency (radio frequency), the lower–spin-state protons can absorb the energy and "jump" to the higher spin state.

A **resonance condition** is a dynamic equilibrium that is established between the lower- and higher-energy spin states. This point can be detected and related to the frequency of energy that is absorbed and is necessary for resonance to occur.

2. THE NMR SPECTRUM

In NMR, a substance to be studied is placed in the presence of electromagnetic radiation and a magnetic field. The electromagnetic radiation is kept at a constant frequency and the strength of the magnetic field is varied. At some value of the magnetic field strength, the energy required to

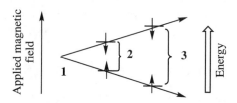

Figure 12-6. Energy states associated with nuclear magnetic resonance (NMR).

"flip" the proton matches the energy of the radiation. At this point, absorption occurs and a signal is observed, which is recorded on an NMR tracing. The **NMR tracing is a plot of absorption of radiation (*y*-axis) versus magnetic field (*x*-axis).**

The frequency at which the proton comes into resonance depends on the chemical and magnetic environment surrounding it, which includes the electron density at the proton and the presence of other, nearby protons. Each proton, or each set of equivalent protons, has a slightly different environment from every other set of protons. The various electronic environments of each kind of proton account for the varied positions of the signals on an NMR tracing.

Four important aspects of an NMR tracing are as follows:

1. **The number of signals** gives information on how many different "kinds" of protons there are in a molecule.
2. The **positions of the signals** tells about the electronic environment of each kind of proton.
3. The **intensities of the signals** reveal the number of each kind of proton.
4. The **splitting of a signal** into several peaks tells about the environment of a proton with respect to other, nearby protons.

3. THE NMR MACHINE

A sample to be studied is placed in a magnetic field and irradiated with a specific frequency of electromagnetic radiation. This frequency usually is at radio frequency. The magnetic field strength is then varied. When it reaches the correct strength, the radio frequency energy is absorbed. The machine can detect this absorbance and records the frequency at which the absorbance occurs.

If electronegative groups are attached to the proton or are on an adjacent carbon, the electron density about that proton is pulled away. This **deshielding** of the proton, causing it to "feel" more of the applied magnetic field, increases the energy separation between the two spin states. In addition, the radio frequency energy needed to bring the proton into resonance is higher.

A standard compound is used for comparison with the compound being studied. Tetramethyl silane (TMS) is often chosen as the internal standard. The methyl protons of this compound are highly shielded, resulting in a relatively small energy separation between the two spin states. Thus, a relatively low radio frequency energy is needed to bring the protons into resonance.

The TMS standard is arbitrarily assigned 0 hertz (Hz) and the rest of the frequencies are reported as the number of hertz away from the TMS signal. Often, the resonances are reported as δ or parts per million (ppm).

$$\delta \, (\text{ppm}) = \frac{[\text{Hz away from TMS}]}{\begin{array}{c}\text{Radiofrequency of}\\ \text{the machine}\end{array}} \times 10^6$$

Thus, all NMR machines, regardless of the radio frequency used, report the same δ-values for identical compounds. The spectrum that results indicates the frequencies at which resonances occurred, and offers important information about the protons in the sample.

4. SHIELDING AND THE CHEMICAL SHIFT

Electrons circulating about the proton itself generate a magnetic field. This field is aligned in such a way that it opposes the applied field. Thus, the field felt by the proton is diminished and the

TABLE 12-2. Chemical Shift Values (δ):

R—CH$_3$	0.8–1.0	R—O—CH$_3$	3.3–3.9
R—CH$_2$—R	1.2–1.4	R—CO—CH$_3$ (ketone)	2.1–2.6
Ph—CH$_3$	2.2–2.5	RCHO (aldehyde)	9.5–9.6
R—CH$_2$—Cl	3.6–3.8	Ph—H	6.0–9.5
R—CH$_2$—Br	3.4–3.6	RC≡CH	2.5–3.1
R—CH$_2$—I	3.1–3.3	R$_2$C=CHR	5.2–5.7

proton is said to be **shielded.** If the circulation of electrons (especially π-electrons) induces a field that reinforces the applied field, the proton is said to be **deshielded.**

Compared to a "naked" proton, a shielded proton requires a higher applied magnetic field strength to provide the particular effective field strength at which absorption occurs. This means that **shielding shifts the absorption "upfield"** (to the right on the NMR tracing). On the other hand, **deshielding shifts the absorption "downfield."** These shifts in position of NMR absorptions arising from shielding and deshielding by electrons are called **chemical shifts** (Table 12-2). Each bond type has a characteristic chemical shift value.

Consider an electronegative group near a proton. Electronegative groups deshield a proton, allowing the proton to "feel" more of the applied magnetic field. A larger energy separation between the spin states results (higher-frequency radio energy needed for resonance to occur). The frequency at which resonance occurs is then shifted to a higher frequency relative to the TMS standard (higher δ, downfield shift). **The more electronegative the adjacent group, the further downfield the signal.** Note that few protons are more highly shielded than TMS; everything else is more deshielded, and resonance occurs downfield, at a specific δ away from TMS.

The molecules that contain aromatic rings resonate far downfield. Electronegativity is not the explanation; benzene, for example, is not very electronegative. Rather, the downfield signal of these molecules is caused by π-electron clouds. The applied magnetic field induces a magnetic field in the π-electron cloud that **deshields the protons on the ring** (Figure 12-7). This effect also occurs with protons on double- and triple-bonded carbons, but not to the same extent.

5. SPIN-SPIN SPLITTING

Along with the chemical environment about a proton, the magnetic environment about a proton also plays a large role. **Protons can feel the different magnetic environments created by hydrogens on adjacent carbons.** If a certain proton is next to a carbon with three protons, the

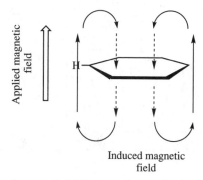

Applied magnetic field

Induced magnetic field

Figure 12-7. Induction of a magnetic field in the π-electron cloud resulting from an applied magnetic field.

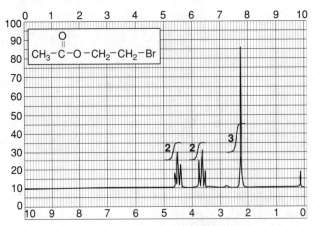

Figure 12-8. A sample NMR tracing.

signal of that proton splits into a quartet. This splitting is attributable to the four ways that three protons of two different spins can be arranged: all three protons with spins aligned with applied magnetic field; two with and one against; one with and two against; and three protons with spins aligned against the applied magnetic field.

If the adjacent carbon has only two protons, the signal is split into a triplet. If the proton is flanked on both sides by carbons with protons, the splitting is more complicated and forms a multiplet. **In general, a set of n equivalent protons splits an NMR signal into n + 1 peaks** (e.g., with three adjacent equivalent protons, the signal is split into a quartet).

Equivalent protons do not split each other. An example is the protons of a methyl group, which are chemically and magnetically equivalent. Another example is $Cl-CH_2-CH_2-Cl$ (1,2-dichloroethane); only one signal is generated for the methylene groups.

Figure 12-8 is an NMR tracing for the molecule $CH_3-CO-OCH_2CH_2Br$, which includes two triplets and a singlet. The rise of the line above the signals is the measure of the area underneath the peak. This value is proportional to the number of protons responsible for the signal.

The integrations indicate that two protons are responsible for each of the triplets and that three protons are responsible for the singlet. The singlet is at 2 ppm, so it must be the methyl next to the carbonyl. The triplets are the methylenes, with the one further downfield belonging to the CH_2 attached to the more electronegative oxygen.

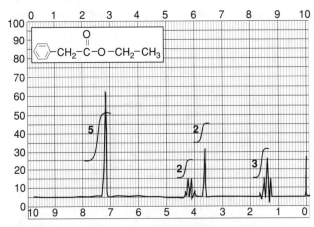

Figure 12-9. A sample NMR tracing.

Figure 12-9 is an NMR tracing for the molecule $Ph-CH_2-CO-OCH_2CH_3$. The combination of a triplet (integrating to three) and a quartet (two protons) suggests an ethyl group. The triplet belongs to the methyl and the downfield quartet belongs to the $-O-CH_2-$methylene. The singlet integrating to two protons is thus the benzyl $-CH_2-$, and the tallest singlet belongs to the aromatic protons.

In summary, it is more important to understand some basic theory about this complex procedure and the principles behind what the tracings show than to be concerned with determining precise chemical structure from NMR tracings.

Appendix: Nomenclature

omenclature

I. General Rules for all Compounds

- **Know the names of the functional groups**, both prefixes and suffixes.
- **Find the longest carbon chain containing the "highest-priority" functional group.** The order in decreasing priority (but not in all cases) follows:
    1. Carboxylic acids and derivatives (esters, amides, nitriles, acid halides, nitriles, anhydrides, and so on)
    2. Aldehydes
    3. Ketones
    4. Alcohols
    5. Amines
    6. Alkynes
    7. Alkenes
    8. Halogens and alkyl groups
- **Number** from the end of the carbon chain, which places the highest-priority functional group at the lowest numeric value.
- **List substituents alphabetically.** Use the suffix for the highest-priority functional group and the prefix for the other functional groups (substituents). **Note:** sec-, iso-, and tert- are placed in alphabetical order, but di-, tri-, and tetra- are not.

II. Rules for Specific Compounds

A. ALKANES AND ALKYL HALIDES

1. **Memorize** prefixes for carbon chains (meth-, eth-, prop-, but-, and so on)—one of the few things in organic chemistry that you need to memorize.
2. **Identify** the longest continuous carbon chain, which determines the prefix. The **suffix for all alkanes is -ane.**
3. **Number** from the chain starting at the end closest to the first branch point or halogen.
    - If substituents are equal distances from the opposite ends, start numbering from the substituent that is first in alphabetical order.
4. If there is more than one of a particular substituent, use di-, tri-, tetra-, and so on before the name of the substituent. (Note: This substituent does not count in the alphabetical order.)
5. **List** the substituents in alphabetical order as shown in Figure A-1. Do not use spaces, just hyphens.
6. **Complex substituents** are named starting at the atom attached to the chain. The names of these substituents are in parentheses.
7. **Common names**

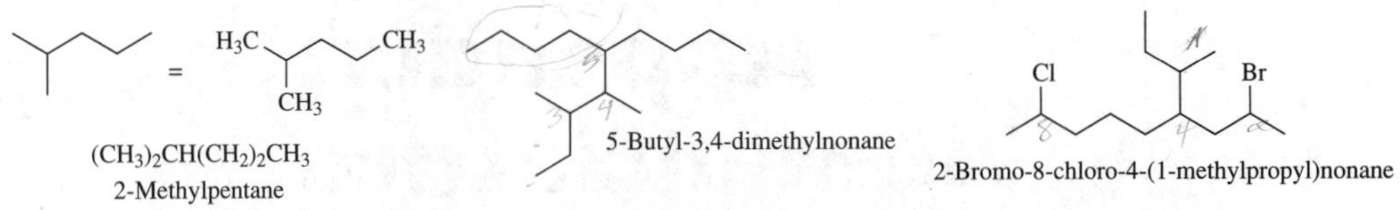

(CH₃)₂CH(CH₂)₂CH₃ — $(CH_3)_2CH(CH_2)_2CH_3$
2-Methylpentane

5-Butyl-3,4-dimethylnonane

2-Bromo-8-chloro-4-(1-methylpropyl)nonane

Figure A-1. Examples of alkane and alkyl halide nomenclature.

3-Carbon

iso-Propyl

4-Carbon

sec-Butyl   *iso*-Butyl   *tert*-Butyl

5-Carbon

iso-Pentyl   *neo*-Pentyl   *tert*-Pentyl

Figure A-2. Common alkane names.

These names are not official names of the International Union of Pure and Applied Chemistry (IUPAC), but are commonplace in many compounds (Figure A-2).

8. **Cyclic alkanes** (Figure A-3)
   a. The base name is the cycloalkane unit.
   b. The same rules apply for naming and ordering of substituents as in the linear case.
      (1) Number substituents so they total to the smallest number.
      (2) Number alphabetically, if possible.

B. ALKENES

1. **Identify** the longest carbon chain containing the double bond, which determines the prefix. The **suffix is -ene.**
2. **Number** from the end closest to the double bond.
   • Multiple double bonds are numbered and the appropriate prefixes (e.g., di-, tri-, and tetra-) are applied (Figure A-4).
3. **List** substituents in alphabetical order using prefixes.
4. **Cis, trans.** These terms identify the isomers.
   a. *cis* alkenes have the larger substituents on the same side.
   b. *trans* alkenes have the larger substituents across the double bond.
5. **E/Z**
   a. *cis* and *trans* are sometimes ambiguous, so IUPAC uses E (opposite) and Z (same or "Zame") designations (see Figure A-4).

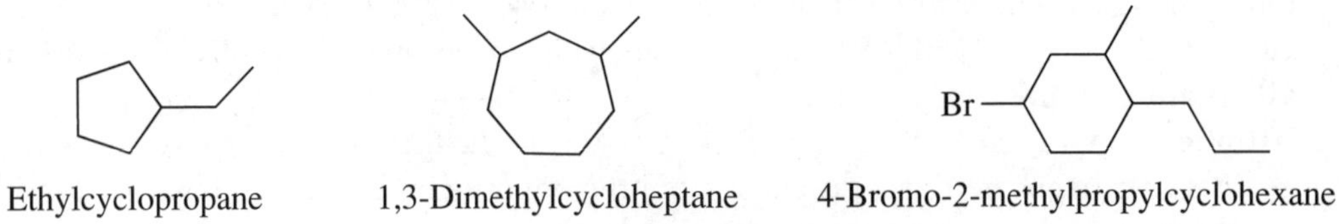

Ethylcyclopropane   1,3-Dimethylcycloheptane   4-Bromo-2-methylpropylcyclohexane

Figure A-3. Examples of cyclic alkane nomenclature.

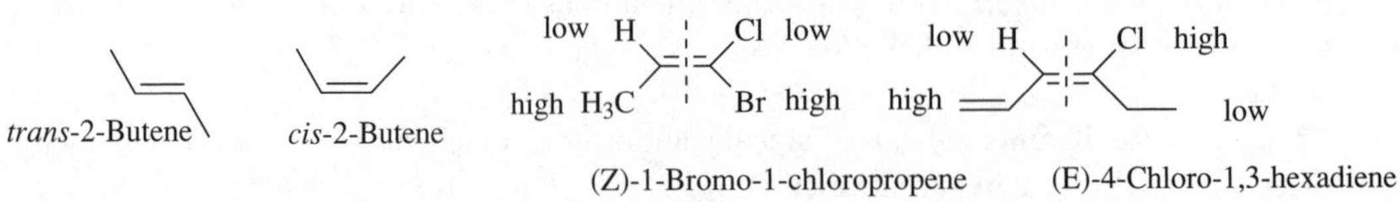

trans-2-Butene   *cis*-2-Butene

low H   Cl low
high H₃C   Br high

(Z)-1-Bromo-1-chloropropene

low H   Cl high
high   low

(E)-4-Chloro-1,3-hexadiene

Figure A-4. Examples of alkene nomenclature.

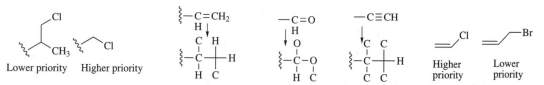

Figure A-5. Branch point and multiple bond considerations in alkene nomenclature.

  b. To designate, **prioritize the sides of each carbon of the double bond.** High priorities on the same side are Z and on the opposite side are E.

6. **Priorities** (also apply for *R/S* designation)
  a. Look at the atom attached to the double bond. Rank by atomic number.
  b. In the case of a tie, go to the next atom and rank by atomic number.
  c. Look for a point of difference such as a branch point. If there still is a tie, go to the next atom. Keep going until a higher-ranking atom is found.
  d. Multiple bonds are tricky. They are counted as if they were the same number of singly bonded atoms: that is, a $CH–CH_2$ double bond counts as four carbons and three hydrogens (Figure A-5).
  e. Remember to look for rankings according to atomic number and not total weight of the substituent group.

7. **Common names**
  a. Mono substituted ethylenes are called vinyl (vinyl chloride or 1-chloroethene in Figure A-5).
  b. When a methylene (CH_2) group is between the double bond and the substituent, the common name is allyl (allyl bromide or 3-bromopropene in Figure A-4). The allyl term is also used in reference to carbocations and radicals.

C.   ALKYNES

1. The **rules** are the same as for alkenes, except **the suffix is -yne.**
2. **Number** from the end closest to the triple bond.

D.   BENZENE COMPOUNDS

1. In the absence of other functional groups, benzene is the base name.
2. **List** the substituents alphabetically and **number** the substituents so that the lowest possible numbers are used.
3. **Ortho, meta,** and **para** can be used to describe disubstituted benzene rings (Figure A-6).
4. **Common names**
  a. When rings are substituents, the names shown in Figure A-7 are used.
  b. Be familiar with the common names in Figure A-8. For each of the common names, substitute the given group for X in the structure. Of the names listed, the benzene carbon attached to the **X group is number 1** in the numbering scheme.

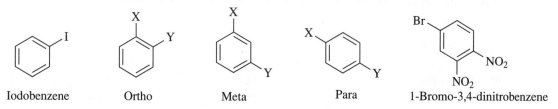

Iodobenzene   Ortho   Meta   Para   1-Bromo-3,4-dinitrobenzene

Figure A-6. Nomenclature of benzene compounds.

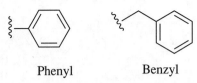

Phenyl   Benzyl

Figure A-7. Phenyl and benzyl groups.

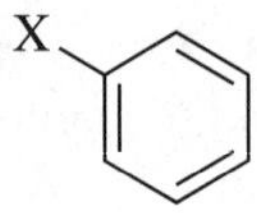

Figure A-8. Common names for benzene compounds (see text).

E. ALCOHOLS

1. The naming of these compounds is similar to alkanes.
   a. **Prefix** is based on the largest carbon chain containing the OH.
   b. **Suffix** is **-ol.**
2. **Numbering** begins at the end closest to the hydroxy group or amino group.
3. **Common names**
   a. Primary (1°), secondary (2°), and tertiary (3°) alcohols are represented in Figure A-9.
   b. Iso-, sec-, and tert- are also used to describe branching.

F. ETHERS

These compounds are named by two different methods.

1. **Simple molecules:** List the alkyl groups in increasing size and add the word *ether*.
2. **More complex molecules:** Use the alkyloxyalkane method. The smaller group has the **-oxy** suffix (Figure A-10).

G. KETONES

1. **Identify** the longest carbon chain with the carbonyl group.
2. **Number** the chain starting at the end closest to the carbonyl.
   • Use the standard prefix for the carbon chain and **-one** as the suffix.

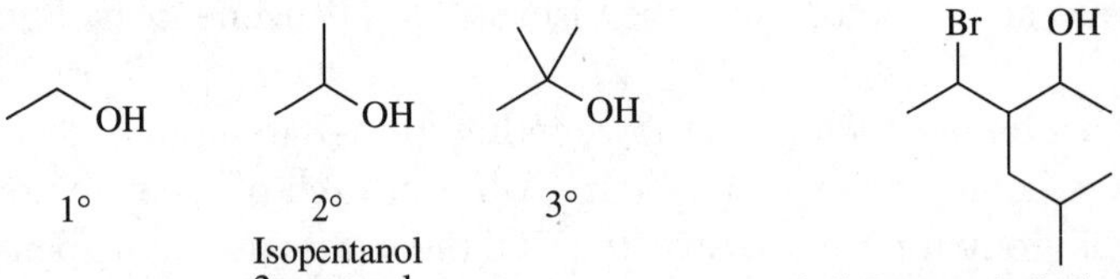

OH	OH	OH	Br OH
1°	2°	3°	
	Isopentanol		3-(1-Bromoethyl)-5-methyl-2-hexanol
	2-propanol		

Figure A-9. Examples of primary, secondary, and tertiary alcohols (left), and a branched alcohol (right).

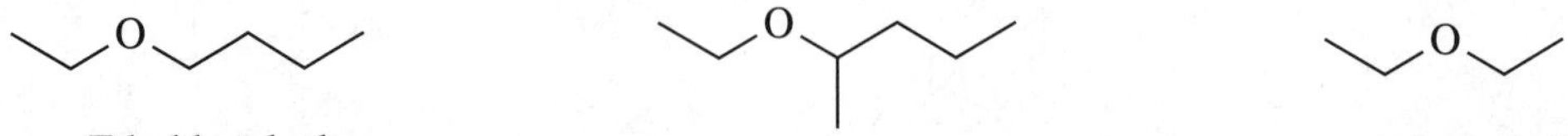

Ethyl butyl ether
or ethoxybutane

2-Ethoxypentane

Diethyl ether
or ethoxyethane

Figure A-10. Nomenclature of ethers.

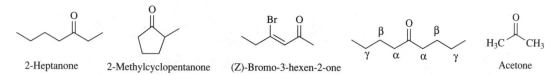

| 2-Heptanone | 2-Methylcyclopentanone | (Z)-Bromo-3-hexen-2-one | | Acetone |

Figure A-11. Nomenclature of ketones.

3. **List** the substituents in alphabetical order (Figure A-11).
4. **Greek letters** are used to describe positions relative to the carbonyl (for all carbonyl compounds).
5. **Remember the common name acet** (as in acetone), which involves the $CH_3CO–$ group.

H. ALDEHYDES, CARBOXYLIC ACIDS, AND ACID CHLORIDES

1. **Identify** the longest carbon chain containing the aldehyde or acid.
2. **Number** the chain starting with the carbonyl carbon.
   a. Use the standard prefix for the carbon chain.
   b. For **aldehydes,** the suffix is **-al.**
   c. For **carboxylic acids,** the suffix is **-ic acid.**
   d. They can also be named (in more complicated cases) as **-carboxylic acid.**
3. **List** the substituents in alphabetical order (Figure A-12).
4. Remember that acet involves a $CH_3CO–$ (acetic acid resembles acetone, except for the hydroxy group of acetic acid).
5. **Acid chlorides** are the same as carboxylic acids, except the suffix is **-yl halide** or **-carbonyl halide.**
6. Also related are **nitriles** ($–C{\equiv}N$), named the same way but the suffix is **-nitrile.**

I. ESTERS

1. **Identify** the alcohol section (connected to oxygen) and the carboxylic acid section (connected to the carbonyl) [Figure A-13].
2. **Name** the alcohol alkyl group.
3. **Finish** the name using alkyl group on the acid side and the suffix **-oate.**
4. **If substituents are attached,** number the alcohol region starting with the carbon connected to the oxygen. On the acid side, start numbering at the carbonyl carbon.

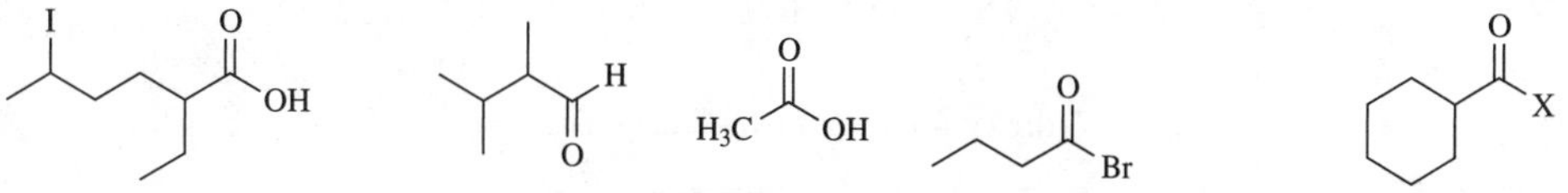

| 2-Ethyl-5-iodohexanoic acid | 2,3-Dimethylbutanal | Acetic acid | Butanoyl bromide | X=OH Cyclohexanecarboxylic acid |
| | | | | X=Cl Cyclohexanecarbonyl chloride |

Figure A-12. Examples of aldehyde, carboxylic acid, and acid chloride nomenclature.

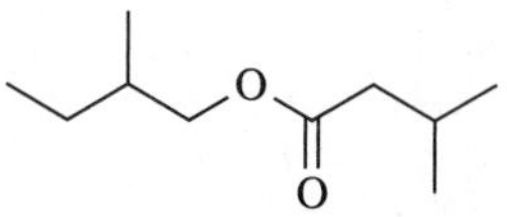

2-(Methyl)butyl 3-methylbutanoate

Figure A-13. An example of the naming of an ester compound.

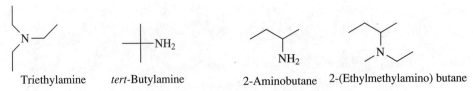

Triethylamine   *tert*-Butylamine   2-Aminobutane   2-(Ethylmethylamino) butane

Figure A-14.   Examples of amine nomenclature.

J.   AMINES

1.   In general, amines are named by listing the alkyl substituents and ending with **-amine.**
2.   For branched primary amines, the prefix **amino-** is applied, along with the position number on the chain.
3.   More complex amines are named by using a combined prefix that uses the largest alkyl group as the base name (Figure A-14).

K.   USE OF PREFIXES

1.   **Prefixes** are used for substituents (which are lower in priority compared with the functional group on which the name is based).
2.   A rough order, which may be different when applied to ring systems, follows. **Prefixes are in bold** and the list is ordered from high to low priority (high dictates the base name).
   a.   Carboxylic acids: **carboxy-** and derivatives (esters, **alkoxyl-** and **carbonyl-**; amides, **amido-**; nitriles, **cyano-**; acid chlorides)
   b.   Aldehydes: **formyl-**
   c.   Ketones: **oxo-**
   d.   Alcohols: **hydroxy-**
   e.   Amines: **amino-**
   f.   Alkynes
   g.   Alkenes
   h.   Halogens and alkyl groups
3.   Figure A-15 is an example showing how to name amides. The base name is the longest chain containing the amide, with the numbering starting at the carbonyl carbon. *N*- is used to describe groups attached to the nitrogen.

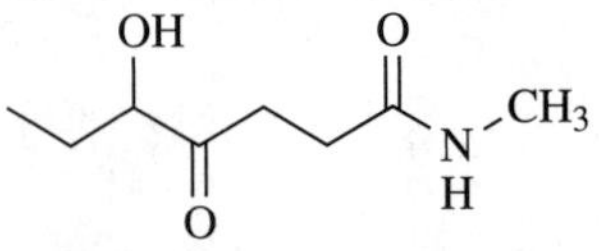

5-Hydroxy-4-oxo-*N*-methylheptamide

Figure A-15.   An example of amide nomenclature.

Section I: General Concepts and Hydrocarbon Chemistry

Structure of Covalently Bonded Molecules

1. Which Lewis structure represents ClO^-?

A. $:\ddot{Cl}-\ddot{O}\cdot$ C. $\cdot\ddot{Cl}-\ddot{O}\cdot$

B. $\cdot\ddot{Cl}-\ddot{O}:$ D. $:\ddot{Cl}-\ddot{O}:$

2. Which Lewis structure represents $SO_4{}^{2-}$ incorporating a neutral sulfur atom?

A. $:\ddot{O}-\overset{\displaystyle :\ddot{O}:}{\underset{\displaystyle :\ddot{O}:}{S}}-\ddot{O}:$ C. $:\ddot{O}-\overset{\displaystyle :\ddot{O}:}{\underset{\displaystyle :\ddot{O}:}{S}}=\ddot{O}$

B. $:\ddot{O}-\overset{\displaystyle :\ddot{O}:}{\underset{\displaystyle :\ddot{O}:}{S}}-\ddot{O}:$ D. $\ddot{O}=\overset{\displaystyle :\ddot{O}:}{\underset{\displaystyle :\ddot{O}:}{S}}=\ddot{O}$

3. Which Lewis structure BEST represents H_2CO?

A. $\underset{\displaystyle H}{H\!:\!C\!::\!\ddot{O}\cdot}$ C. $:\ddot{O}\!:\!C\!:\!\overset{\displaystyle H}{H}$

B. $H\!:\!C\!:\!H\,\ddot{O}$ D. $H\!:\!\overset{\displaystyle H}{C}\!::\!\ddot{O}$

4. Which Lewis structure represents the molecule with the atoms in the order Cl, N, N, Cl?

A. $:\ddot{Cl}\!:\!N\!:\!N\!:\!\ddot{Cl}:$ C. $:\ddot{Cl}\!:\!N\!::\!N\!:\!\ddot{Cl}\,:$

B. $:\ddot{Cl}\!:\!\ddot{N}\!::\!N\!:\!\ddot{Cl}\,:$ D. $:\ddot{Cl}\!:\!N\!\cdot\!N\!:\!\ddot{Cl}:$

5. The Lewis structure for CH_3COOH contains:

A. two double bonds.
B. five singly bonded atoms.
C. four pairs of unbonded electrons.
D. unbonded electrons on carbon.

6. What is the electron dot formula for H_2S?

A. $:H\!:\!H\!:\!\ddot{S}$ C. $:\ddot{S}\!:\!H\!:\!H$

B. $H\!:\!\ddot{S}\!:\!H$ D. $H\!:\!\ddot{S}\!:\!H$

7. Without considering hyperconjugation, how many resonance structures can be drawn for the $SO_3{}^{2-}$ ion?

A. One   C. Three
B. Two   D. Four

8. Without considering hyperconjugation, how many resonance structures can be drawn for the following molecule?

$$CH_2=CH-CH^+$$

A. One   C. Three
B. Two   D. Four

9. Without considering hyperconjugation, how many resonance structures can be drawn for 1,3-butadiene?

A. One   C. Three
B. Two   D. Four

10. Which resonance structures for 1,3,5-hexatriene is the most stable?

A. $^+CH_2CH=CH-CH=CH-CH^-$
B. $CH_2=CH-CH=CH-CH=CH_2$
C. $^-CH_2-CH=CH-CH=CH-CH_2{}^+$
D. All have equal stability.

11. Which statement is true?

A. Resonance hybrids are inherently unstable.
B. Resonance hybrids are more stable than any individual resonance form.
C. Resonance hybrids are averages of all resonance forms resembling the less stable forms.
D. Resonance hybrids are averages of all resonance forms resembling the more stable forms.

12. Without considering hyperconjugation, how many uncharged resonance structures can be drawn for pyridine?

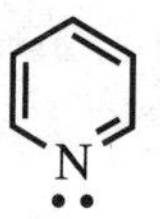

A. One C. Three
B. Two D. Four

13. Without considering hyperconjugation, how many resonance structures can be drawn for the following compound?

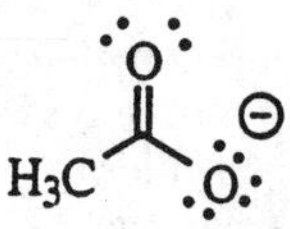

A. One C. Three
B. Two D. Four

14. Without considering hyperconjugation, how many resonance structures can be drawn for the following compound?

$$H_2N-\overset{O}{\underset{\overset{\oplus}{NH_2}}{\overset{\|}{C}}}-NH-CH_2-CH_2-CH_2OH$$

A. One C. Three
B. Two D. Four

15. The atomic orbitals for carbon of methane are considered:

A. nonhybridized orbitals.
B. sp hybridized.
C. sp^2 hybridized.
D. sp^3 hybridized.

16. The compound BF_3 is an example of an sp^2 hybridized molecule. How many valence electrons are in its central atom and how many unhybridized p orbitals?

A. 2, 1 C. 3, 1
B. 2, 2 D. 3, 2

17. Which statement about sigma (σ) bonds is NOT true?

A. They are circularly symmetrical in cross section when viewed along the bond axis.
B. Electron density is contained between their atomic nuclei.
C. They form when the 1s orbital of hydrogen overlaps hybrid orbitals of carbon.
D. None of the above.

18. Beryllium has an atomic number of 4. The molecule BeH_2 is expected to be:

A. unhybridized.
B. sp^2 hybridized.
C. sp^3 hybridized.
D. sp hybridized.

19. What is the bond angle between the hybrid orbitals in methane?

A. $180°$
B. $120°$
C. $115.5°$
D. $109.5°$

20. Which geometry of molecules that are sp hybridized reflects their hybrid orbitals?

A. Trigonal pyramidal
B. Bent
C. Tetrahedral
D. Linear

21. The hydrocarbon molecule ethyne has a total of:

A. one σ bond, two π bonds.
B. two σ bonds, four π bonds.
C. three σ bonds, two π bonds.
D. one σ bond, four π bonds.

22. Double bonds consist of:

A. one σ bond, one π bond.
B. two σ bonds, one π bond.
C. one σ bond, two π bonds.
D. two σ bonds, two π bonds.

Questions 23–25 refer to the following diagram:

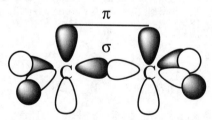

23. Which hydrocarbon is BEST shown by the orbital framework?

A. Ethane
B. Ethene
C. Ethyne
D. Ethanal

24. Which angle is closest to the one between hybrid orbitals in the diagrammed molecule?

   A. 109.5°   C. 135°
   B. 120°   D. 150°

25. The geometry of this molecule is BEST described as:

   A. linear.
   B. trigonal planar.
   C. tetrahedral.
   D. trigonal bipyramidal.

26. What is the bond length of a carbon–carbon double bond?

   A. 1.20 Å   C. 1.54 Å
   B. 1.34 Å   D. 1.68 Å

27. The geometry of the compound SF_6 is BEST described as:

   A. octahedral.
   B. tetrahedral.
   C. trigonal pyramidal.
   D. trigonal bipyramidal.

28. What is the three-dimensional shape of ammonia, NH_3, which is an sp^3 hybridized molecule?

   A. Bent
   B. Tetrahedral
   C. Trigonal pyramidal
   D. Trigonal bipyramidal

29. The hybridization and geometry for compound PCl_5 are

   A. sp^3, tetrahedral.
   B. sp^3, trigonal bipyramidal.
   C. sp^3d, trigonal bipyramidal.
   D. sp^3d, trigonal pyramidal.

30. The hybridization and geometry for the oxygen of H_2O are

   A. sp^2, planar.
   B. sp^3, tetrahedral.
   C. sp^2, bent.
   D. sp^3, bent.

31. The length of carbon-to-carbon bonds in rank of *increasing* bond length is:

   A. triple, double, single.
   B. single, double, triple.
   C. single, triple, double.
   D. triple, single, double.

32. Which molecule has a dipole moment equal to zero?

   A. NF_3 C. BF_3
   B. NH_3 D. CH_2Cl_2

33. Which molecule has a nonzero dipole moment?

   A. Cl_2 C. CCl_4
   B. CO_2 D. $CHCl_3$

34. Which molecule has the greatest dipole moment?

   A. CH_3Cl C. CH_3F
   B. CH_3Br D. CH_3I

35. Which molecule has the greatest dipole moment?

   A. CH_4 C. CH_2Cl_2
   B. CH_3Cl D. $CHCl_3$

36. Ethylamine, $CH_3CH_2NH_2$:

   A. contains polar covalent bonds.
   B. has a net dipole moment.
   C. contains nitrogen with a partial negative charge.
   D. all of the above.

37. The dipole moment of a molecule depends on:

   A. the magnitude of the separated partial positive and negative charges.
   B. the distance between separated charges.
   C. both of the above.
   D. neither of the above.

SOLUTIONS

Structure of Covalently Bonded Molecules

1. **D** The net charge of the molecule is -1. Therefore, when the valence electrons are placed around each atom, one atom must have one extra electron placed around it. Cl has seven valence electrons and O has six. Placing the extra electron around the Cl atom would fill its octet, and no covalent bond could form. The extra electron must be placed on the O atom.

    :C̈l· ·Ö: ⟶ :C̈l:Ö:

2. **D** Again in this question, the molecule has a net charge. The central atom is sulfur, which normally has six valence electrons. The net charge of -2 suggests that two extra electrons must be placed around each O. For this problem, place six electrons around S. Here O has six valence electrons; therefore, place six electrons around each O, which leads to two double bonds, as shown.

    extra electron
    4x Ö:
    :S̈: → O=S=O
    2 extra electrons
    extra electron

3. **D** Oxygen likes to form two bonds. Because it binds to carbon, and carbon has two Hs bound to it, structure D is the best Lewis structure.

4. **C** Nitrogen likes to form three bonds. Because nitrogen has five valence electrons, a single pair of electrons must remain unbonded. Chlorine has seven valence electrons and room to bond only one more to complete its octet. Only structure C allows each nitrogen to have three bonds, and the chlorine to have one bond (sharing an e^- to complete its octet).

5. **C** Because oxygen likes to form two bonds and have two pairs of unbonded electrons per oxygen, two oxygen atoms in this Lewis structure would result in four pairs of unbonded electrons.

    H_3C Ö Ö–H

6. **D** Sulfur, the central atom, has six valence electrons. Two of the six electrons are involved with bonds to hydrogen; therefore, two electron pairs are unbonded. The geometry of the molecule is called *bent*, which is very similar to the structure of H_2O. Remember that S and O are in the same column on the periodic table.

    S
    H   H

7. **C** Resonance structures are generated by moving unbonded pairs of electrons toward single bonds, thereby generating double bonds. Existing double bonds are broken, forming single bonds and an electron pair. The diagrams that follow show the movement of the three resonance structures for the given ion. Single bonds can also be broken (hyperconjugation), but the position of the atoms must be maintained. Resonance forms involve the movement of electrons, not nuclei.

    :O=S—O ⟷ :O—S—O
    :O:⊖   :O:

8. **B** One pair of electrons in the double bond can shift to the adjacent single-bond position, thereby neutralizing the carbocation (positive charge) there. Note that a carbocation is generated on the carbon on the left because its

electron was used to form the π bond on the other side.

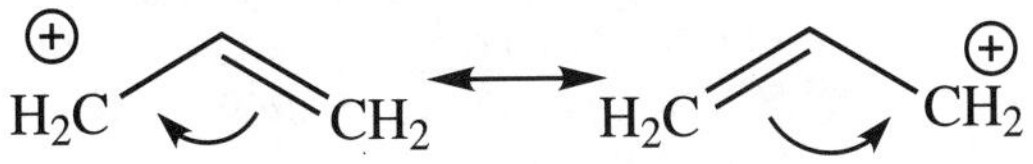

9. **C**   The diagram shows how three resonance structures can be generated. *Cis* and *trans* forms are isomers. Resonance forms do not involve the movement of the atoms, only their electrons.

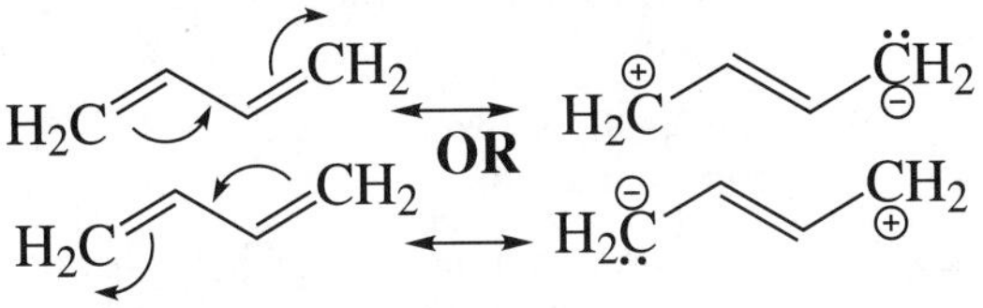

10. **B**   Choices A and C are resonance structures that contribute to the overall resonance hybrid (or average of all

resonance forms). However, these forms are charged and are therefore less stable than their uncharged counterparts.

11. **D**   Resonance is an application of molecular orbital theory and suggests that electrons are not really localized in a bond between two atoms of a molecule, but instead are spread throughout the molecule. Resonance hybrids are the averages of these delocalized resonance forms, with preference given to the more stable resonance forms.

12. **B**   The diagram shows how two resonance structures can be generated by moving π bond electrons.

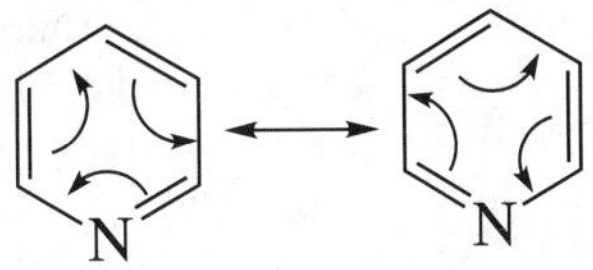

13. **C**   An electron pair from the π bond moves to the oxygen, giving it a negative charge and the carbon a positive charge. A lone pair from the other oxygen then forms a π bond with the carbon, resulting in the other carbonyl compound. A double bond

does not form between the methyl group and the carbonyl carbon because such a bond would require the loss of a hydrogen. Resonance structures should involve only the movement of electrons.

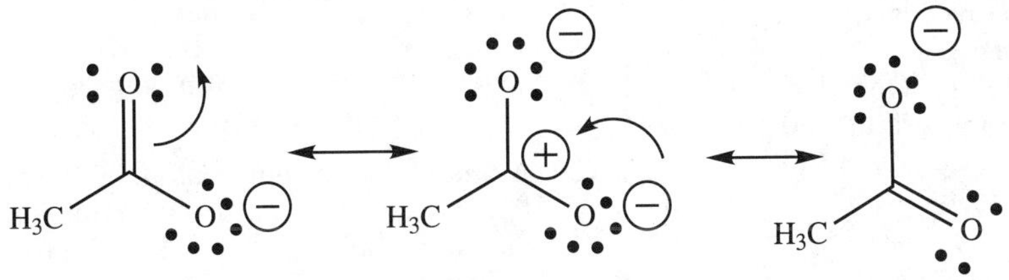

14. **D**   An electron pair from the nitrogen can form a π bond, and the electrons from the already existing π bond can then be used to neutralize the positive charge. This situation creates a positive charge on the atom that gave up the electron. The four resonance structures that can be drawn are:

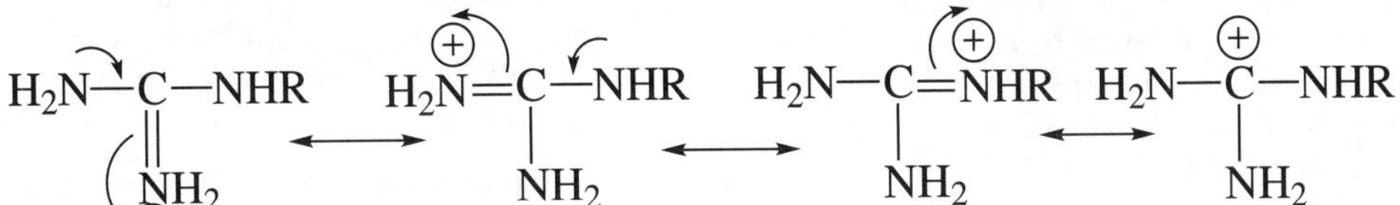

$$R = -CH_2CH_2CH_2OH$$

15. **D** Carbon has four valence electrons, with two electrons in the 2s subshell and two in the 2p. Hybridization maximizes the overlap of atomic orbitals when forming bonds (molecular orbitals). The 2s orbital is mixed with the three 2p orbitals, resulting in the formation of four $2sp^3$ orbitals. One electron is placed in each hybrid orbital, each orbital is 109.5° from any other and each forms a tetrahedral geometry. This arrangement places the electrons as far from one another as possible.

16. **C** Boron has three valence e^-; two in the 2s and one in the 2p. The 2s orbital is mixed with two of the three 2p orbitals, resulting in three $2sp^2$ orbitals and one unhybridized 2p orbital. One electron is placed in each hybrid orbital, which is planar and 120° from any other and can bond with F. The 2p orbital is empty and can accept electrons. This feature makes many boron-containing molecules good Lewis acids. This arrangement places the electrons as far from one another as possible.

17. **D** All the facts given in choices A–C are true.

18. **D** Be has two valence e^- filling the 2s orbitals. The 2s orbital is mixed with one of the three 2p orbitals, resulting in two 2sp hybrid orbitals and two unhybridized 2p orbitals. One electron is placed in each hybrid orbital and oriented 180°. This arrangement places the electrons as far from one another as possible.

19. **D** Discussed in question 15.

20. **D** The sp orbitals are 180° from one another, and the molecule assumes a linear geometry.

21. **C** σ-bonds (end-on-end overlap of orbitals) occur owing to the overlap of sp hybridized orbitals in this molecule and the overlap of sp hybrid orbitals with the 1s orbitals of hydrogen. The π bonds (sideways overlap of orbitals) form between the unhybridized p orbitals. Note a total of three σ and two π bonds.

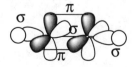

22. **A** The diagram shows that one σ and one π bond would form. Double bonds involve sp^2 hybridization, in which one unhybridized p orbital from each atom interacts to form one π bond.

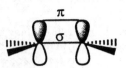

23. **B** Each carbon has three sp^2 hybridized orbitals. Each must be a double-bond–containing compound, namely ethene.

24. **B** The sp^2 hybridized orbitals are 120° from one another.

25. **B** The bond geometry of sp^2 hybridized molecules is trigonal planar. The sp hybridized molecules are linear; sp^3 hybridized molecules are either tetrahedral (four substituents), trigonal pyramidal (three substituents, one e^- pair), or bent (two substituents, two e^- pairs). The sp^3d is trigonal bipyramidal when there are five substituents or four substituents and one e^- pair, and sp^3d^2 is often octahedral with six substituents.

26. **B** The C–C bond is 1.54Å. The C=C bond is 1.34Å and the C≡C bond is 120Å.

27. **A** The sulfur has six substituents (see solution to question 25).

28. **C** Ammonia has a central atom of N, with five valence e^-. Because ammonia has only three substituents, it has one free e^- pair, which classifies the geometry as trigonal pyramidal (see solution to question 25).

29. **C** P has five valence electrons. The five substituents make this situation a classic sp^3d trigonal bipyramidal molecule.

30. **D** The oxygen in water forms one bond with each hydrogen and has two pairs of unbound electrons. The overall hybridization is sp^3, and the geometry is classified as bent (see solution to question 25).

31. **A** See solution to question 26. Multiple bonds result in shorter and stronger bonds.

32. **C**   In choices A and B, the compounds contain nitrogen, which has a pair of unbonded electrons and a trigonal bipyramidal shape. This arrangement causes a net dipole moment, as can be seen in the diagram. Choice C involves boron, which has only three valence e^- and its sp^2 is hybridized. Note that the vectors cancel because of the symmetry. Choice D also has a net dipole moment because of its tetrahedral shape and its two electronegative substituents, which create a polar vector.

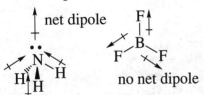

net dipole

no net dipole

33. **D**   Choices A–C are all symmetric and the dipole vectors cancel. Only choice D describes a molecule with a nonzero dipole moment.

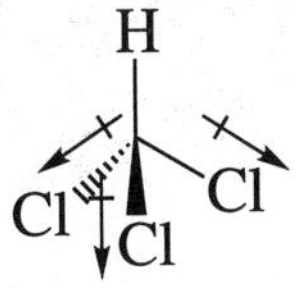

34. **C**   Fluorine has the greatest electronegativity and pulls electrons away from the alkyl group to the greatest extent. This combination gives fluoromethane the greatest dipole moment.

35. **B**   Because Cl is an electron-withdrawing group, one might expect that the more Cl atoms present, the greater the pull of electrons from the C. Although this description is true, the dipole moment is also concerned with the vector direction of electron attraction. Because of vector canceling in choices C and D, choice B actually has the greatest dipole moment.

36. **D**   Because N is more electronegative than C, the covalent bond between them can be considered polar. The molecule has a dipole moment, and the nitrogen contains a partial negative charge.

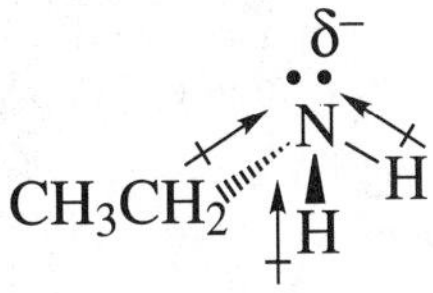

$$\delta^-$$
$$CH_3CH_2 \quad N \quad H$$
$$H$$

37. **C**   Choices A and B both describe the dipole moment.

Section I: General Concepts and Hydrocarbon Chemistry

Stereochemistry of Covalently Bonded Molecules

1. Ethanol (CH_3CH_2OH) and dimethyl ether ($(CH_3)_2O$) are BEST considered:

   A. structural isomers.
   B. stereoisomers.
   C. enantiomers.
   D. diastereomers.

2. *Cis* and *trans* isomers of alkenes are BEST considered:

   A. structural isomers.
   B. meso compounds.
   C. enantiomers.
   D. identical structures.

3. Which pair of compounds are the stereoisomer of one another?

   I. Pair of structural isomers
   II. Pair of enantiomers
   III. Pair of diastereomers
   IV. Pair of geometric isomers

   A. I, II, and III
   B. II, III, and IV
   C. I, III, and IV
   D. II and IV

4. Which is NOT true about chiral or asymmetric carbon centers?

   A. They must contain four different substituents.
   B. An example of a molecule containing a chiral carbon is 2-propanol.
   C. Any molecule containing only one chiral carbon can have a pair of enantiomers.
   D. A molecule may contain more than two chiral centers.

5. The molecule that has a chiral carbon is:

   A. 1-chloropropane.
   B. 1-chloro-2-methylpropane.
   C. 2-bromobutane.
   D. 3-chloropentane.

6. The BEST name for the following compound is:

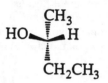

   A. 2-butanol.
   B. D-2-butanol.
   C. (S)-2-butanol.
   D. (R)-2-butanol.

7. Consider (R)- and (S)-2-butanol. Which physical property distinguishes the two compounds?

   A. Melting point
   B. Solubility in common solvents
   C. Infrared spectrum
   D. Rotation of plane-polarized light

8. The number of degrees that a plane of polarization is rotated as it passes through a solution of an enantiomer depends on:

   I. concentration of the enantiomer.
   II. length of the polarimeter tube.
   III. whether R and S enantiomers of a compound are present.

   A. I
   B. II
   C. I and II
   D. I, II, and III

9. Which of the following represents a racemic mixture?

   A. 75% (R)-2-butanol, 25% (S)-2-butanol.
   B. 25% (R)-2-butanol, 75% (S)-2-butanol.
   C. 50% (R)-2-butanol, 50% (S)-2-butanol.
   D. none of the above.

Questions 10–12 refer to the structure of the following carbohydrate molecule:

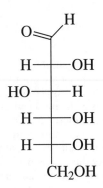

10. How many chiral carbons does this structure contain?

   A. One   **C.** Five
   B. Two   **D.** Four

11. Considering the number of chiral carbons, what is the maximum number of stereoisomers for this compound?

   A. 2   **C.** 8
   B. 4   **D.** 16

12. This carbohydrate molecule is BEST considered:

   A. optically active.
   B. optically inactive.
   C. a meso compound.
   D. two of the above.

Questions 13–17 refer to the following three compounds:

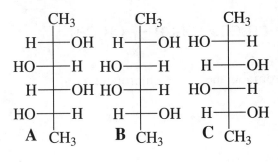

13. Which structures are enantiomers?

   I.   A and B
   II. B and C
   III. A and C

   A. I
   B. II
   C. III
   D. I and II

14. Which structures are diastereomers?

   I.   A and B
   II. B and C
   III. A and C

   A. I
   B. II
   C. III
   D. I and II

15. Which compound(s) is (are) optically active?

   A. A
   B. B
   C. C
   D. A and C

16. Which pair(s) of compounds has (have) different physical properties?

   A. A and B
   B. B and C
   C. A and C
   D. A and B, B and C

17. Which pair(s) of structures is (are) stereoisomers?

   A. A and B
   B. A and B, A and C
   C. A and B, B and C
   D. A and B, B and C, A and C

SOLUTIONS

Stereochemistry of Covalently Bonded Molecules

1. **A** Isomers are different compounds that have the same molecular formula. The two main categories of isomers are structural isomers and stereoisomers. Structural isomers differ in the connections of their atoms in space. Draw the structures of ethanol and dimethyl ether. Note that they have different orders of their atoms. Therefore, they are structural isomers.

2. **C** Stereoisomers are compounds that have the same connections but different orientations in space. For alkenes, *cis* and *trans* isomers differ in their orientation about the carbon–carbon double bond. Therefore, they are stereoisomers. The two types of stereoisomers are enantiomers and diastereomers. Enantiomers are nonsuperimposable compounds that are mirror images of one another (e.g., your hands are mirror images of one another and they cannot be superimposed).

Diastereomers are nonsuperimposable compounds that are not mirror images of one another. Meso compounds are molecules with chiral centers that have a plane of symmetry and are not optically active. *Cis* and *trans* isomers are not enantiomers and are not superimposable; therefore, they are considered diastereomers of one another. *Cis* and *trans* isomers are sometimes called geometric isomers. In this situation, they are not mirror images and are not superimposable. Because the term geometric isomer describes orientation of molecules in space, the geometric isomer is a type of stereoisomer.

3. **D** Explained in the solution to question 2

4. **B** The statements made in choices A, C, and D are true. Choice B is not true because the molecule does not contain a carbon with four different substituents.

5. **C** The structures for the compounds are shown. Look for the compound that has four different substituents attached to carbon. This situation occurs only in choice C.

H   H   H   H
| | | |
Cl—CH₂CH₃   Cl-CH₂—CH₃   H₃C—CH₂CH₃   CH₃CH₂—CH₂CH₃
| | | |
H   CH₃   Br   Cl
1-chloropropane   1-chloro-2-   2-bromobutane   3-chloropentane
                         methylpropane

6. **D** The nomenclature for compounds that have enantiomers gives both enantiomers a separate and distinct name—either R or S. Start by ranking the substituents by atomic number. If a tie occurs, compare the next atom of each substituent until one has a higher atomic number. Place the lowest priority group to the back, and look at the order of the other three in terms of rank. R is a clockwise and S is a counterclockwise ordering by priority.

7. **D** Enantiomers have identical physical properties except for their behavior toward plane-polarized light. When a plane of polarized light is passed through a chiral compound, the plane is rotated. Pure enantiomers rotate the plane-polarized light in equal but opposite directions. Therefore, equal mixtures of two enantiomers provide no net rotation of plane-polarized light; in other words, the solution is optically inactive. *Meso* compounds, for the same reason, are optically inactive. In one molecule, they contain two chiral centers with the same substituents but opposite configurations. Because they are always in solution at the same concentration, each cancels the optical activity of the other.

8. **D** A polarimeter is an apparatus for measuring the rotation of plane-polarized light as it passes through a solution of an optically active compound. Polarimeters contain both a light source that emits light on various planes of orientation and a polarizer (filter) that allows only a single plane of polarized light to pass through it. After passing through the polarizer, the plane-polarized light passes into the polarimeter tube, which contains the optically active substance. The longer the tube, the greater the rotation of the plane-polarized light, because it contacts the optically active substance for a longer time. The rotation of the plane-polarized light also depends on the concentration of the substance and whether enantiomers of the substance are present (enantiomers cancel each other's optical activity). The relationship between the specific rotation of light $[\alpha]$, observed rotation α, concentration of the optically active solution (c), and length of the polarimeter tube (l) is:

$$[\alpha] = \frac{\alpha}{(c)(l)}$$

$[\alpha]$ = specific rotation
α = observed rotation
c = concentration of solution g/mL
l = length of tube (dm)

9. **C** Racemic mixtures are equimolar mixtures of two enantiomers. These mixtures are optically inactive because the optical activity of the two enantiomers are opposite to one another. When the rotation of plane-polarized light for two compounds is opposite, the optical activity of the mixture becomes zero. This leads to an optically inactive mixture.

10. **D** The four carbons shown in the diagram have four different substituents and are therefore considered chiral. All of the carbons must have single bonds.

11. **D** The maximum number of stereoisomers for a compound is 2^n, where n = number of chiral centers.

12. **A** The compound is optically active because it does not have a plane of symmetry (meso compound) and it contains chiral centers.

13. **C** Enantiomers are nonsuperimposable mirror images.

14. **D** Compounds A and B as well as B and C are pairs that are not superimposable and are not mirror images of one another.

15. **D** Compound B is meso because it contains a plane of symmetry between carbons 3 and 4. Meso compounds are optically inactive. The other compounds are optically active.

16. **D** Diastereomers have different physical properties.

17. **D** Stereoisomers include pairs of enantiomers and diastereomers.

118

Section I: General Concepts and Hydrocarbon Chemistry

Alkanes

1. What is the correct IUPAC name for the following compound?

   H₃C $\diagdown$ $\diagup$ $\diagdown$ $\diagup$ CH₃
   CH₂CH₃

   A. 2-Ethylpentane
   B. 4-Ethylpentane
   C. 3-Methylhexane
   D. 4-Methylhexane

2. What is the correct IUPAC name for the following compound?

   CH₃
   H₃C $\diagdown$ $\diagup$ $\diagdown$ $\diagup$ CH₃
   CH₃ CH₂CH₃

   A. 4-Methyl-2,5-dimethylheptane
   B. 4-*Sec*-butyl-2-methylhexane
   C. 4-Ethyl-2,5-dimethylheptane
   D. 2-Methyl-4-*sec*-butylhexane

3. What is the correct IUPAC name for the following compound?

   H₃C $\diagdown$ $\diagup$ $\diagdown$ Br
   CH₃

   A. Isopropyl bromide
   B. 1-Bromo-2-methylpropane
   C. *Sec*-butyl bromide
   D. 3-Bromo-2-methylpropane

4. Which octane molecule has the highest boiling point?

   A. Branched octane
   B. Unbranched octane
   C. Cyclo-octane
   D. All have the same boiling point

5. How do the boiling points of butane, propane, and ethane compare?

   A. The boiling point of ethane is greatest.
   B. The boiling point of butane is greatest.
   C. The boiling point of propane is greatest.
   D. There is no relationship among these three alkanes and their boiling points.

6. How does the melting point of propane and ethane compare?

   A. The melting point of propane is greater.
   B. The melting point of ethane is greater.
   C. The melting points are within 5°C of one another.
   D. No relationship exists among these three alkanes and their boiling points.

7. As the molecular weight of alkanes increases, how do the boiling point and melting point change?

   A. Boiling point increases; melting point increases.
   B. Boiling point increases; melting point decreases.
   C. Boiling point decreases; melting point decreases.
   D. Boiling point increases; melting point increases sequentially for alkanes with over four carbons.

8. The branching of alkanes that produces symmetrical structures:

   A. raises the boiling point.
   B. lowers the melting point.
   C. raises the melting point.
   D. is both A and C.

9. The sum of the coefficients in the balanced reaction for the combustion of butane is:

   A. 16.5.   C. 12.5.
   B. 13.   D. 10.

10. Which gas is NOT expected to have ring strain?

    A. Cyclopropane
    B. Cyclobutane
    C. Cycloheptane
    D. None of the above

11. The most stable conformation of cyclohexane is the:

    A. Hayworth form.
    B. sawhorse form.
    C. Newman form.
    D. chair form.

12. What percentage of cyclohexane molecules is estimated to be in the boat form at any given moment?

   A. Over 99%
   B. Between 90% and 99%
   C. Approximately 50%
   D. Less than 1%

13. Methylcyclohexane can have two conformations in the chair form. Which form is the most stable?

   A. The methyl group in an axial position
   B. The methyl group in an equatorial position
   C. Both conformations equally stable
   D. Both conformations unstable because the boat form is favored for substituted cycloalkanes

14. Which dimethylcyclohexane is the most stable?

A. 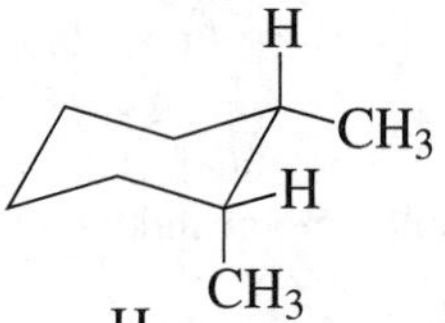
C.

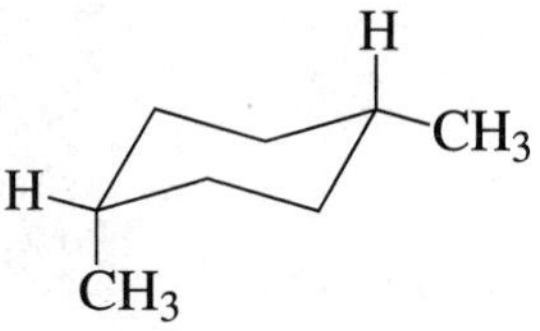

B. 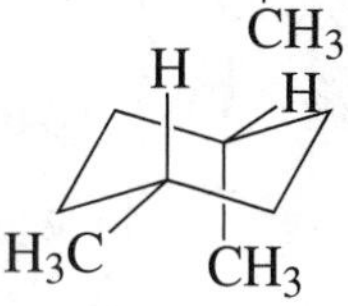
D.

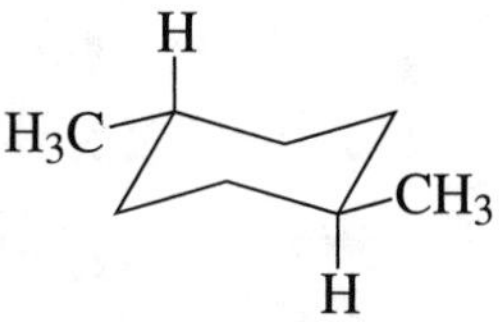

15. In the following substituted cyclohexane, the hydroxyl groups are in which position on the ring?

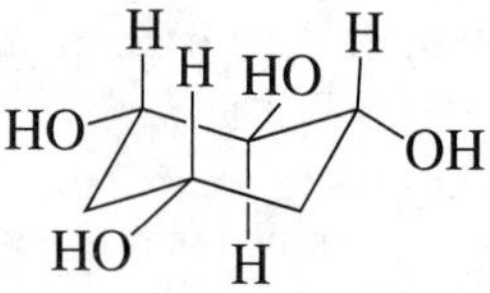

   A. Axial
   B. Equatorial
   C. Both axial and equatorial
   D. Not enough information to determine

16. Which chair form BEST depicts the trimethylcyclohexane in the Hayworth projection?

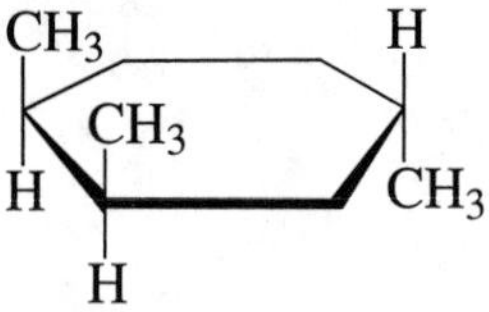

A. 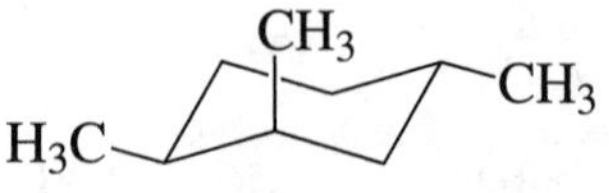
B. 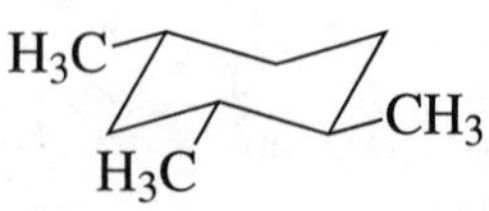
C.
D.

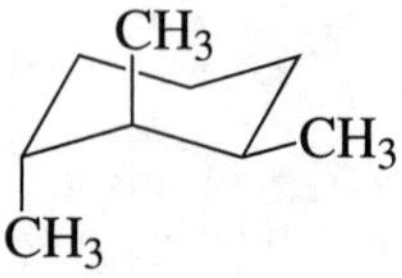

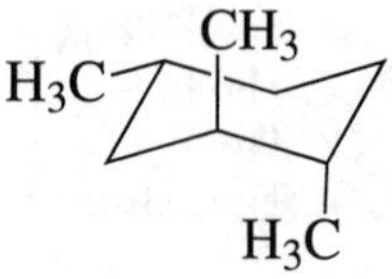

17. The correct IUPAC name for the following compound is:

- **A.** *cis*-1,4-dimethylcyclohexane.
- **B.** *trans*-1,4-dimethylcyclohexane.
- **C.** 1,4-dimethylcyclohexane.
- **D.** all of the above.

18. Which halogen does NOT appreciably react with methane in a free-radical substitution reaction?

- **A.** Iodide
- **B.** Chlorine
- **C.** Fluorine
- **D.** Bromine

19. In the chlorination alkanes, the first step in which chlorine free radicals are produced is called:

- **A.** initiation.
- **B.** activation.
- **C.** propagation.
- **D.** deactivation.

20. Chlorine free radicals react with methane by:

- **A.** donating their free-radical electron to methane.
- **B.** abstracting a hydrogen atom from methane, and producing HCl and a methyl radical.
- **C.** forming a carbanion intermediate that rapidly dissociates to produce chloromethane.
- **D.** donating their free-radical electron to methane to from chloromethane.

21. In the termination step of free radical substitution reactions, the alkyl free radical reacts with:

- **A.** another alkyl free radical.
- **B.** free electrons.
- **C.** halogen free radicals.
- **D.** A and C

22. Why is the halogenation of alkenes considered a chain reaction?

- **A.** It occurs quickly.
- **B.** It occurs without the generation of intermediates.
- **C.** Each step generates the reactive intermediate that causes the next step to occur.
- **D.** The reaction allows long chains of halogenated alkanes to be formed.

23. Which alkyl free radical is the most stable?

- **A.** Methyl
- **B.** Primary
- **C.** Secondary
- **D.** Tertiary

24. Of the three free radicals shown, which is the most stable?

$$\text{I} \qquad \text{II} \qquad \text{III}$$

- **A.** I
- **B.** II
- **C.** III
- **D.** All are equally stable.

25. What factor accounts for the stability of the free radicals that have resonance structures?

- **A.** A greater ability to delocalize electron charge
- **B.** A greater mass over which to spread electron charge
- **C.** Both
- **D.** Neither

26. What product is formed in the free-radical bromination of methane?

- **A.** Bromomethane
- **B.** Dibromomethane
- **C.** Tribromomethane
- **D.** All of the above

SOLUTIONS

Alkanes

1. **C** Identify the longest carbon chain. Start numbering from the side of the chain that has a substituent closest to its end. In this example, the substituent is a methyl group in the 3 position.

$$H_3C \underset{5}{\overset{6}{\diagdown}} \underset{}{\overset{4}{\diagup}} \underset{\underset{2 \quad 1}{CH_2CH_3}}{\overset{3}{\diagdown}} CH_3$$

2. **C** Start numbering from the left side of the molecule, as this approach results in a lower number to the first methyl group than numbering from the right. The longest chain has seven carbons; therefore, the prefix hept will be used.

3. **B** Numbering begins with the carbon connected to the Br substituent. The methyl group is attached to the number 2 carbon of propane. The common name for this compound is isobutyl bromide.

4. **C** Cyclic alkanes tend to have higher boiling points than noncyclic alkanes. Branched alkanes tend to have lower boiling points than unbranched alkanes because branched alkanes tend to have lower van der Waals attraction forces between their molecules compared to unbranched alkanes.

5. **B** The boiling point in alkanes increases with increasing molecular weight because of enhanced van der Waals interactions. However, if branching occurs, the boiling point is lower.

6. **B** The unbranched alkanes do not show a smooth increase in melting points with increasing molecular weight. An alternation occurs when progressing from methane (mp = −183°C) to ethane

(mp = −172°C) to propane (mp = −188°C) to butane (mp = −138°C). The melting points increase with increasing molecular weight, after butane.

7. **D** Discussed in solutions to questions 6 and 8.

8. **D** In general, branching lowers boiling points of alkanes. However, branching that produces highly symmetrical molecules results in higher boiling points and melting points.

9. **A** The combustion of alkanes produces carbon dioxide and water. The balanced reaction is:

$$C_4H_{10} + 13/2 \ O_2 \rightarrow 4 \ CO_2 + 5 \ H_2O$$

10. **D** Remember that sp^3 hybridization favors a bond angle of 109.5°. Rings are not always planar as they look on paper, but can pucker to accommodate the more favorable bond angles. The only cycloalkane with no ring strain is cyclohexane, because it can pucker and form the chain conformation in which each carbon is tetrahedral and each bond is in a staggered conformation. Cyclopropane has the most ring strain, and cyclobutane has almost as much ring strain. Among the choices given, cycloheptane has the lowest ring strain, followed by cyclo-octane.

11. **D** The chair is the most stable conformation that cyclohexane can assume. The boat form is another conformation, but it is higher in energy because of eclipsing strain and flagpole interactions. Hayworth, sawhorse, and Newman are projections used to picture organic molecules. They do not specifically refer to cyclohexane.

chair   boat   Hayworth Newman sawhorse

12. **D** Over 99% of cyclohexane molecules are in the energetically favorable chair conformation at any given time.

13. **B** Any group has considerably more room when it occupies an equatorial position. There is less potential repulsion for the methyl group with axial hydrogens (1, 3 diaxial interactions), and therefore greater stability.

14. **D** The best choice is the one in which both methyl groups are in the equatorial positions. This position gives maximum stability because the two methyl groups tend to repel one another, and equatorial positioning as shown in D maximally separates the methyl groups.

15. **B** The following diagram shows a cyclohexane in the chair conformation and identifies the equatorial and axial positions.

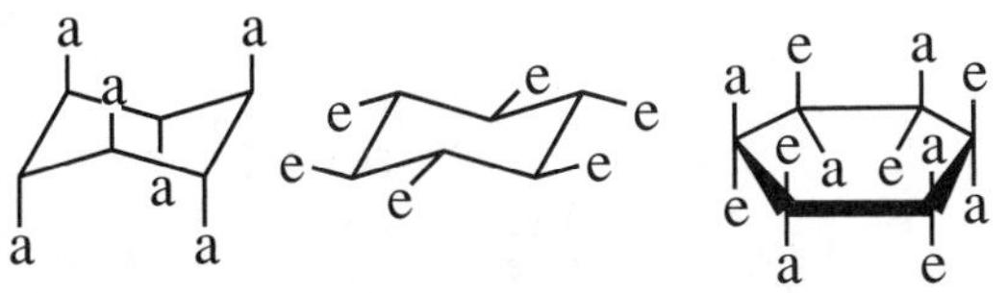

16. **A** Practice translating the Hayworth projection to the chair projection and back, as shown in the solution for question 15.

17. **B** Because the methyl groups are on the opposite side of the ring, the compound is known as *trans*. Start numbering from a carbon on the ring to which a methyl group is attached.

18. **A** Iodide radicals are highly unreactive because of their relatively low electronegativity when compared with the other halogens. This low negativity is due to the "largeness" of the atom and the relatively long distance from the outer shell to the nucleus. Therefore, iodide radicals react very poorly in free-radical substitution.

19. **A** Initiation is the net production of free radicals. Propagation is the transfer of the free radical electron to activate an alkyl group (no net increase in free radicals). Termination is the net decrease in free radicals.

Initiation $\qquad X_2 \xrightarrow[\text{or } \Delta]{hv} X\text{-}X \longrightarrow 2\,X^\bullet$

Propagation $\qquad X^\bullet \; H\text{-}H_2C\text{-}R \longrightarrow R\text{-}CH_2 + HX$

$\qquad\qquad R\text{-}\overset{\bullet}{C}H_2 \; X\text{-}X \longrightarrow R\text{-}CXH_2 + X^\bullet$

Termination $\qquad 2\,R\text{-}\overset{\bullet}{C}H_2 \longrightarrow RCH_2CH_2R$

$\qquad\qquad R\text{-}\overset{\bullet}{C}H_2 + X^\bullet \longrightarrow R\text{-}CXH_2$

$\qquad\qquad 2\,X^\bullet \longrightarrow X_2$

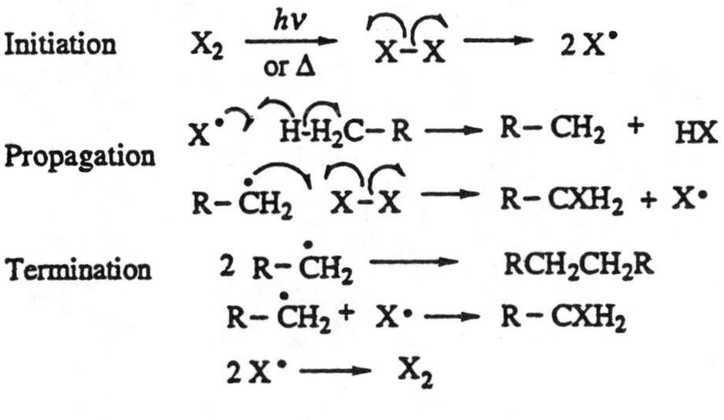

20. **B** Shown in solution to question 19

21. **D** Shown in solution to question 19. Choices A and C are correct.

22. **C** Note that each propagation step of the free radical substitution of alkanes produces a product required for the next step to occur. Adding halogens to alkanes tends to substitute halogens for hydrogens on alkyl groups. This addition leads to halogenation of alkanes, not production of polymers.

23. **D** Benzyl > allyl > tertiary > secondary > primary > methyl. Benzyl and allyl free radicals are stable in that they have the ability to delocalize the free radical electron over resonance structures. The bulky alkyl groups of the tertiary free radical allow it to spread out the free radical electron charge effectively.

24. **B** Described in solution to question 23

25. **A** This question is tricky. Choice A is correct. Choice B suggests that spreading the charge over a greater mass increases stability. Although stability may increase, the question asks why resonance-producing compounds are very stable. Many compounds that have resonance structures have small masses.

26. **D** Free-radical reactions tend to produce mono-, di-, and trisubstituted halogenated alkanes rapidly. It is often difficult to stop a free-radical reaction from producing di- and trisubstituted alkanes.

Section I: General Concepts and Hydrocarbon Chemistry

Alkenes, Substitution and Elimination Reactions

1. The correct IUPAC name for the following compound is:

   H_3C—CH_2CH_3
   H_3C—CH_2CH_3

   A. 2-methyl-3-propyl-2-pentene.
   B. 3-ethyl-2-methyl-2-pentene.
   C. 4-methyl-3-propyl-3-pentene.
   D. 2-methyl-3-ethyl-hex-2-ene.

2. The correct IUPAC name for the following compound is:

   CH_3CH_2—CH_3
   H—CH_2Cl

   A. E-1-chloro-2-methyl-2-pentene.
   B. Z-1-chloro-2-methyl-2-pentene.
   C. cis-1-chloro-3-methyl-3-pentene.
   D. trans-1-chloro-3-methyl-3-pentene.

3. The correct IUPAC name for the following compound is:

   H_3C—CH_3
   H—H

   A. trans-2-butene.
   B. cis-2-butene.
   C. Z-2-butene.
   D. B and C

4. The correct IUPAC name for the following compound is:

   Cl—Cl
   H—Br

   A. cis-1,2-dichloro-1-bromoethene.
   B. Z-1-bromo-1,2-dichloroethene.
   C. trans-1,2-dichloro-1-bromoethene.
   D. E-1-bromo-1,2-dichloroethene.

5. The correct IUPAC name for the following compound is:

   CH_3CH_2—H
   H—CH_3

   A. cis-2-pentene.
   B. trans-2-pentene.
   C. cis-3-pentene.
   D. trans-3-pentene.

6. Which statements about alkenes and alkanes of corresponding chain length is true?

   I. Alkenes have slightly lower melting points than alkanes.
   II. Alkenes have slightly higher melting points than alkanes.
   III. Alkenes have higher boiling points than alkanes.
   IV. Alkenes have lower boiling points than alkanes.

   A. I and III
   B. II and IV
   C. I and IV
   D. II and III

7. In which solvent are alkenes most soluble?

   A. Water
   B. Ethyl alcohol
   C. Carbon tetrachloride
   D. Ammonia

8. Why do trans isomers of alkenes have lower boiling points than the cis isomers?

   A. Trans isomers have better symmetry.
   B. Cis isomers have better symmetry.
   C. Trans isomers are less polar.
   D. Cis isomers are less polar.

9. How do the melting points of trans isomers compare to the cis isomers for alkenes?

   A. Cis isomers have higher melting points.
   B. Trans isomers have higher melting points.
   C. Both have similar melting points.
   D. No consistent trend is observed.

10. How does the dipole moment of *cis*-1,2-difluoroethene compare to that of *trans*-1,2-difluoroethene?

   A. The dipole moment of the *cis* isomer is greater.
   B. The dipole moment of the *trans* isomer is greater.
   C. The dipole moments of the two isomers are the same.
   D. Not enough information is available to determine the relationship.

11. Rank the molecules 1–3 in order of decreasing stability:

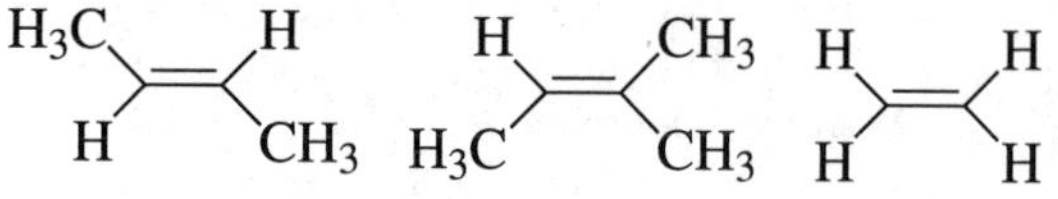

$$H_3C \diagdown \diagup H \qquad H \diagdown \diagup CH_3 \qquad H \diagdown \diagup H$$
$$H \diagup \diagdown CH_3 \qquad H_3C \diagup \diagdown CH_3 \qquad H \diagup \diagdown H$$

   1   **2**   **3**

   A. 3, 1, 2
   B. 3, 2, 1
   C. 1, 2, 3
   D. 2, 1, 3

12. Which statement about the chemistry of carbon–carbon double-bonded compounds is NOT true?

   A. The double bond of alkenes acts to stabilize carbonium ions on adjacent carbons.
   B. The double bond attracts electrophiles.
   C. The double bond acts to stabilize free radicals on adjacent carbons.
   D. The double bond attracts nucleophiles.

13. Rank the following carbocations in order of increasing stability:

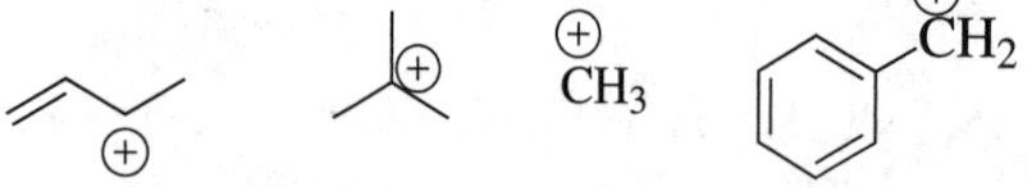

   1   **2**   **3**   **4**

   A. 4, 1, 2, 3
   B. 3, 2, 4, 1
   C. 4, 2, 1, 3
   D. 3, 2, 1, 4

14. In the reaction of ethene and HCl, the H^+ ion acts as the:

   A. nucleophile.
   B. electrophile.
   C. carbanion.
   D. carbonium ion.

Questions 15 and 16 refer to the following reaction:

$$H \diagdown \diagup CH_3 \quad \xrightarrow{\text{HCl}} \quad ?$$
$$H \diagup \diagdown H$$

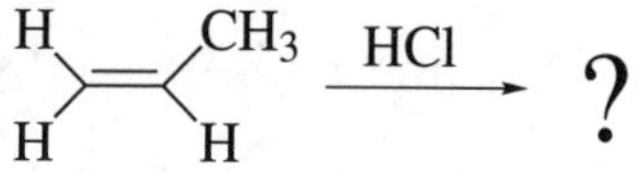

15. Of the intermediates that form, which one predominates?

   A. 1° carbocation
   B. 2° carbocation
   C. 3° carbocation
   D. Alkyl carbocation

16. In this reaction, the Cl^- ion acts as the:

   A. nucleophile.
   B. electrophile.
   C. carbanion.
   D. carbonium ion.

17. In the reaction of H_2O with ethene in the presence of H_2SO_4, which one adds across the double bond first?

   A. H^+
   B. ^-OH
   C. H_2O as an intact molecule
   D. Sulfate ion

18. Why does a H^+ ion attacking a carbon–carbon double bond add to the carbon with the LEAST number of substituents?

   A. The reaction is resonance stabilized.
   B. The hybrid orbital geometry favors this process.
   C. Nucleophiles tend to attack stable centers of negative charge.
   D. A more stable carbocation is generated.

Questions 19–23 refer to the following reaction:

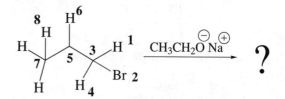

19. The mechanism for this reaction predicts that the base (sodium ethoxide) binds the atom labeled:

   A. 1.   C. 5.
   B. 2.   D. 6.

20. Which combination of positions is most likely to form a double bond?

  A. 1 and 3
  B. 2 and 3
  C. 3 and 5
  D. 5 and 7

21. The rate of this reaction is said to be:

  A. unimolecular.
  B. bimolecular.
  C. trimolecular.
  D. polymolecular.

22. Which one is a major product of this reaction?

  A. Water
  B. HBr
  C. Ethyl alcohol
  D. H_2BrOH

23. What type of intermediate is formed in this reaction?

  A. Free radical
  B. Carbanion
  C. Carbocation
  D. No reaction intermediate

Questions 24–29 refer to the following reaction:

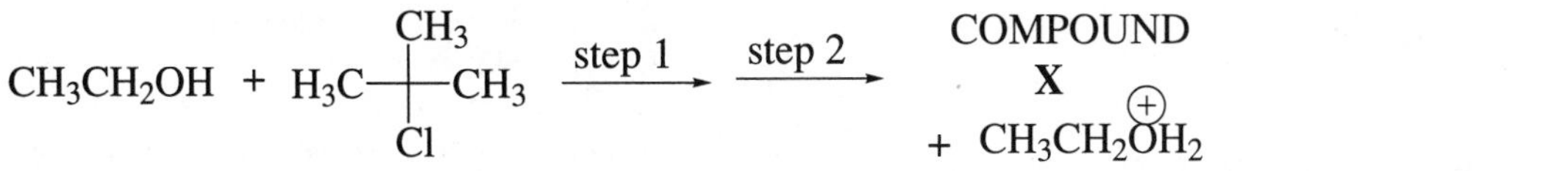

$$CH_3CH_2OH \; + \; H_3C-\overset{\displaystyle CH_3}{\underset{\displaystyle Cl}{\overset{|}{\underset{|}{C}}}}-CH_3 \; \xrightarrow{\text{step 1}} \; \xrightarrow{\text{step 2}} \; \begin{array}{c} \text{COMPOUND} \\ \text{X} \\ + \; CH_3CH_2\overset{+}{O}H_2 \end{array}$$

24. This type of reaction is BEST classified as:

  A. E1.   **C.** E2.
  B. S_N1.   **D.** S_N2.

25. Which step is rate-limiting?

  A. 1
  B. 2
  C. Both
  D. Neither

26. The reaction is BEST classified as:

  A. zero order.
  B. unimolecular.
  C. bimolecular.
  D. trimolecular.

27. The product of step 1 is called a:

  A. transition state.
  B. carbanion.
  C. carbocation.
  D. free radical.

28. In this reaction, ethanol most likely acts as the:

  A. nucleophile.
  B. acid.
  C. base.
  D. carbanion.

29. The correct IUPAC name for compound X is:

  A. *tert*-butyl ethyl ether.
  B. *tert*-butyl alcohol.
  C. 3-methylpropane.
  D. 2-methylpropene.

30. In an S_N1 reaction, the generation of the carbocation occurs in:

  A. the first step.
  B. the second step.
  C. the third step.
  D. none of the steps.

31. The rate of an S_N1 reaction is affected by:

  A. the concentration of the nucleophile.
  B. the identity of the nucleophile.
  C. the geometric orientation of the nucleophile.
  D. none of the above.

32. Which electrophile can readily undergo an S_N1 reaction?

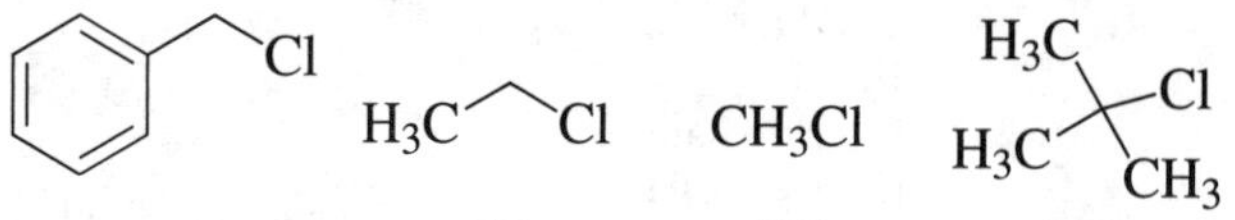

I   II   III   IV

A. I, II, and IV
B. IV
C. I and IV
D. I, II, III, and IV

33. The major product of the reaction between *tert*-butyl chloride and water via an S_N1 mechanism is:

A. 2-butene.
B. 3-butene.
C. 3-butanol.
D. *tert*-butyl alcohol.

Questions 34–38 refer to the following reaction:

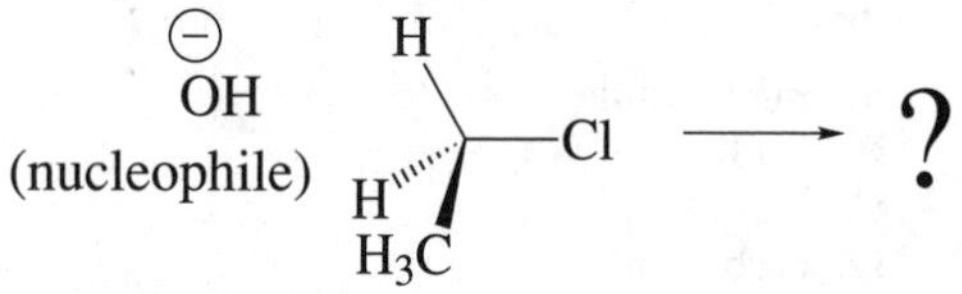

34. What is the principal product of this reaction?

A. Propane   C. Ethene
B. Propanol   D. Ethanol

35. In the mechanism for this reaction, the nucleophile approaches the carbon bearing the Cl atom from the:

A. side bearing the Cl atom.
B. side opposite the Cl atom.
C. side opposite the methyl group.
D. side bearing the methyl group.

36. Which statement is NOT true?

A. If the H's can be differentiated, the primary product of this reaction is an inverted configuration compared to the primary reactant.
B. The rate of this reaction depends on the concentration of alkyl halide and the nucleophile.
C. This reaction proceeds at a rate greater than that of chloromethane and a nucleophile.
D. This reaction has a decreased rate of substitution product formation if strong base is added.

37. What happens when iodine replaces chlorine in the reactant molecules?

A. The reaction rate decreases.
B. The reaction rate increases.
C. The reaction rate stays the same.
D. Not enough information is available to determine the answer.

38. Which action leads to an increased rate of substitution product formation?

A. Adding sterically hindered base
B. Substituting Cl with F
C. Placing additional alkyl groups on the carbon that is bound to the leaving group.
D. None of the above

39. Substitution and elimination reactions compete with one another. Which reaction type predominates when a weak base is applied at room temperature for methyl halides?

A. S_N1 C. E1
B. S_N2 D. E2

40. For 1° alkyl halides, which action does NOT tend to increase the yield of elimination products?

A. Increasing temperatures
B. Adding sterically hindered base
C. Adding CH_3COO^- as a base
D. Adding NH_2^- as a base

41. S_N1 and E1 predominate in reactions involving 3° alkyl halides. Which action tends to shift the product formation toward substitution?

A. Adding a stronger base
B. Adding *tert*-butoxy ion
C. Adding a catalyst
D. Lowering the temperature

128

SOLUTIONS

Alkenes, Substitution and Elimination Reactions

1. **B** Identify the longest carbon chain and begin numbering from the end closest to the double bond. The double bond is between the second and third carbon; therefore, the leftmost methyl group is labeled 1. The compound is then named with the substituents in alphabetic order.

$$\underset{\underset{4\quad 5}{\overset{1}{H_3C}}}{\overset{\overset{1}{H_3C}}{}} \quad \overset{2}{\underset{3}{C=C}} \quad \begin{matrix} CH_2CH_3 \\ CH_2CH_3 \end{matrix}$$

2. **A** The E/Z naming system identifies geometric isomers. Identify the substituent on the double bond that has the highest atomic number. In this problem, the double-bonded carbon on the left has a carbon (ethyl) and a hydrogen group bonded to it. The double-bonded carbon on the right has a carbon (methyl) and a second carbon (from a chloromethyl group). Since all the carbons have the same atomic number, move to the next atom. For the methyl group, the next atom is a hydrogen. For the chloromethyl group, the next atom of highest priority is a chlorine, therefore, this group has a higher priority. Using an analogous process on the other double-bonded carbon (left), the ethyl group is found to have the higher priority. If the two higher priority groups occur on the same side of the double bond, the compound is designated Z. If the higher priority groups occur on opposite sides of the double bond, the compound is designated E. To name this compound, give the position of the double bond (in this situation, 2), and identify the longest carbon chain. This compound is E.

$$\underset{\underset{*}{H}}{\overset{\overset{*}{CH_3CH_2}}{}} \quad C=C \quad \begin{matrix} CH_3 \\ CH_2Cl \end{matrix}$$

* higher priority

3. **D** Since two substituents are present, the *cis/trans* system and E/Z system can be used. When the two substituents are on the same side of the double bond, the compound is designated *cis*. When the two groups are on the opposite side of the double bond, the compound is designated *trans*. Since only two groups are present, the methyl groups have higher priority over the hydrogen substituents by atomic number. Therefore, choices B and C are both correct.

$$\underset{\underset{H}{\overset{\overset{*}{H_3C}}{}}}{} \quad C=C \quad \underset{\underset{H}{}}{\overset{\overset{*}{CH_3}}{}}$$

4. **D** The high priority group on the double-bond carbon on the left is chlorine, whereas on the right it is bromine. Therefore, the compound is in an E configuration; that is, the two high-priority groups are on opposite sides of the double bond. List the substituents in alphabetic order.

5. **B** The two groups attached to the C=C are on opposite sides of the double bond.

6. **A** Alkenes have lower melting points than corresponding alkanes because the alkenes lack regularity of shape. The alkenes have higher boiling points than the corresponding alkanes because the alkenes often are more polar (the C=C bond with draws electrons, and *cis/trans* isomers may contribute to polarity).

7. **C** The hydrocarbons, for example, alkenes, are most soluble in nonpolar solvents. Carbon tetrachloride is nonpolar (individual polarity vectors cancel), and the other solvents listed have some polarity.

8. **C** The *trans* isomers are less polar. The net polarity vectors for *cis* isomers are partially additive, whereas the polarity vectors for *trans* isomers cancel. Molecule symmetry is not the primary reason why this trend exists, however;

symmetry differences occur between *cis/trans* isomers. This symmetry difference relates to the differences in *cis/trans* isomer polarity.

9. **B**   *Trans* isomers tend to have higher melting points, lower boiling points, and lower dipole moments than the corresponding *cis* isomers.

10. **A**   Note that the *cis* isomers tend to have greater dipole moments than the *trans* isomers because of a lack of symmetry and less vector canceling.

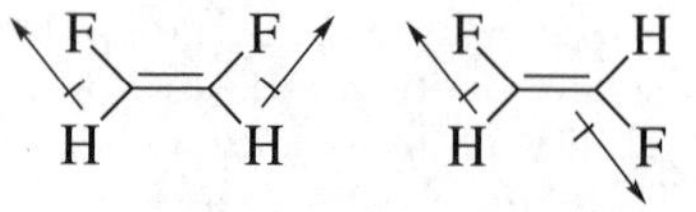

11. **D**   The more alkyl groups around a double bond, the more stable the double bond.

The alkyl groups are electron-releasing and increase C=C bond stability.

12. **D**   The double bond tends to be electron-rich and attract electrophiles (electron-seeking molecules and ions). Because of their electron-rich character and ability to spread out charge, double bonds can stabilize carbonium ions or free radicals on adjacent carbons. This stabilization occurs via resonance in which the p-orbital of the carbonium or free radical is lined up with the π-bond of the double bond.

13. **D**   Carbocations are most stable when they are stabilized by resonance. In addition, the greater the number of alkyl substituents, the greater the stability because of the electron-releasing characteristics of the alkyl substituents.

14. **B**   The following mechanism shows that the electron-rich C=C attacks the electron-deficient proton (H^+) and forms a carbocation intermediate. Because the H^+ sought to gain an electron to neutralize the charge, it acted as an electrophile.

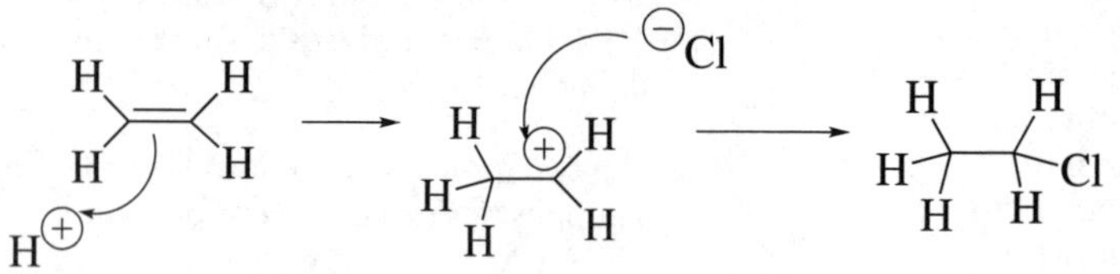

15. **B**   To answer such a question, draw the intermediate. Remember that it is easier to add H^+ to the carbon that has the greater number of H's attached, leading to the most highly substituted carbocation. In this situation, the proton adds to the carbon on the left, resulting in a secondary carbocation. Addition of the proton to the carbon on the right would lead to the less stable (higher energy) primary carbocation (lower population).

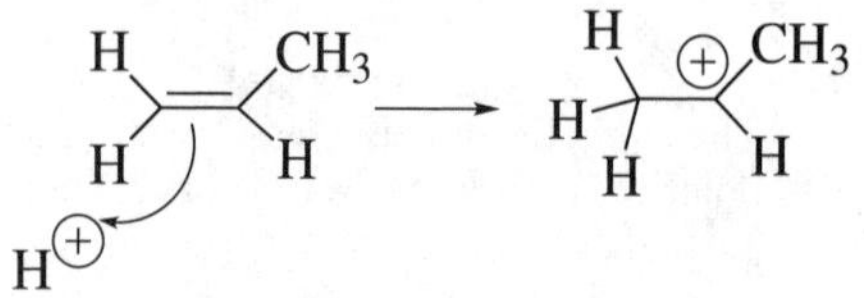

16. **A**   The Cl^- (nucleophile) has an extra electron, and this charge can be neutralized by attacking a positively charged species (electrophile). It attacks the carbocation that is formed from the electrophilic attack of the double bond by H^+ (see the solution to question 12).

17. A The mechanism is similar to the one for the addition of HCl to a double bond (question 12). In this example of a hydration reaction, an acid catalyst is needed to provide a proton in the form of H_3O^+ (hydronium ion) to initiate the addition. Only a catalytic amount is needed, however, because the proton is regenerated.

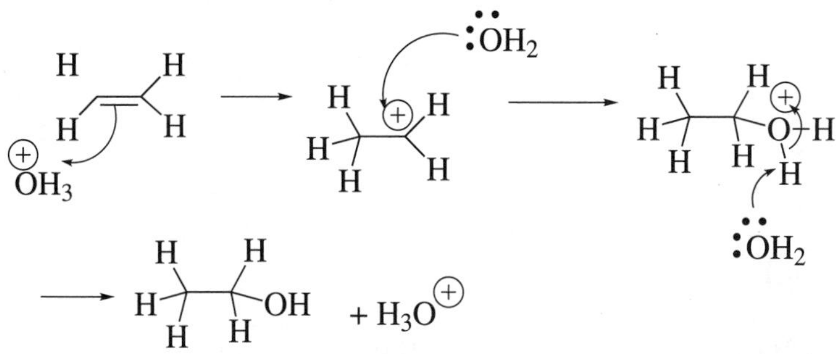

18. D This situation involves Markovnikov's rule, which can be explained by the stability of the carbocation intermediates that are generated. When the proton attacks the double bond, it attaches to the carbonyl with the least number of substituents, therefore generating the more highly substituted (more stable) carbocation.

19. C When a strong base attacks a primary alkyl halide, an E2 reaction can occur. The E2 reaction competes with the corresponding S_N2. In the E2, the base accepts (abstracts) a proton from the β position, leaving behind the electrons and allowing them to form a double bond as the leaving group departs.

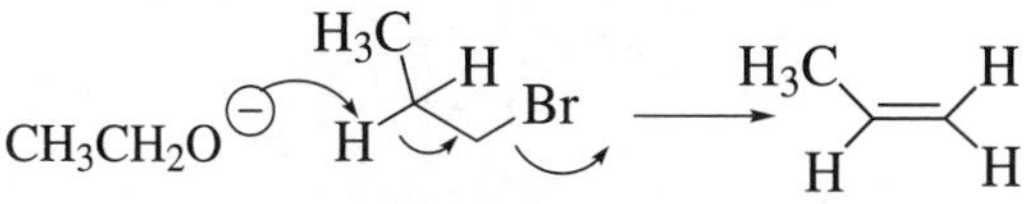

20. C The preceding mechanism shows where the double bond forms. One of the carbons of the double bond is the place where the leaving group was originally attached. Generally, these reactions follow Zaitsev's rule and form the most highly substituted double bond by preference. Exceptions occur if the substrate, the base, or both is highly hindered sterically.

21. B This reaction is known as E2, which stands for bimolecular elimination. In other words, the rate depends on the concentration of both the electrophile and the base. E2 reactions tend to occur with 1° and 2° alkyl halides whereas E1 reactions occur readily with 3° alkyl halides. E1 reactions are called unimolecular because their reaction rates depend only on the concentration of the electrophile.

22. C The strong base, the ethoxide, abstracts a proton from the β carbon, producing ethyl alcohol.

23. D Note that no intermediate is formed. The one-step reaction goes through a transition state (see solution to question 19).

24. **A** The first thing to notice is that this reaction involves a 3° alkyl halide and a weak base. Therefore, this reaction will go through either S_N1 or E1. These two reactions compete such that products from both reactions occur. However, the protonated ethyl alcohol shows that this reaction has undergone E1.

same for both **OR**

E_1 and S_N1

25. **A** The first step in both S_N1 and E1, the carbocation formation, is the rate-limiting step. This step has a higher energy of activation barrier compared with the second step (double-bond formation). For S_N1, the second step is the nucleophilic attack.

26. **B** The E1 mechanism is unimolecular because its rate depends only on the concentration of the alkyl halide. The slowest step is the formation of the carbocation intermediate, which does not depend on the presence of base.

27. **C** The mechanism shows the formation of the carbocation.

28. **C** Ethanol acts as a base by accepting a proton from the substrate. In real situations, the two reactions (S_N1 and E1) are always in competition and product mixtures generally result.

29. **D** See the structure in the reaction for the solution to question 24.

30. **A** See the solution to question 24.

31. **D** Remember that the rate for either S_N1 or E1 mechanisms depends only on the concentration of the electrophile leading to the carbocation.

32. **C** The alkyl halides that react via S_N1 tend to be those that form stable carbocations after the leaving group departs. The stability of carbocations is dictated by resonance and substitution. Therefore, in this question, the benzyl halide (leading to the resonance-stabilized carbocation) and the tertiary alkyl halide can react via

S_N1 or E1. Although 1 cannot undergo E1, other benzyl halides with β protons can. 1° and 2° alkyl halides can react via S_N2 or E2.

33. **D** The mechanism is the same as the one for S_N1 in the solution to question 24 except that in this situation, water is the nucleophile.

34. **D** This is an S_N2 reaction. Primary alkyl halides usually react with nucleophiles by an S_N2 mechanism unless the base is strong and sterically hindered (then the reaction tends to become E2).

35. **B** In S_N2 reactions, the nucleophile approaches on the side opposite the leaving group (backside attack), which is the most electropositive area of the molecule due to the polarity of the carbon-leaving group dipole. Therefore, bulky alkyl groups on the carbon bearing the leaving group make S_N2 less likely to occur. The nucleophile has more trouble approaching the electrophile. This steric hindrance makes it more likely for the nucleophile to act as a base and abstract a proton from the less hindered β carbon, and E2 results.

36. **C** Chloromethane is a methyl halide, which always undergoes S_N2. Furthermore, the reaction proceeds faster than the substitution of a 1° alkyl halide because

of sterics. Choices A and B accurately describe S_N2 whereas choice D is true because strong bases favor the E2 reaction in this situation.

37. **B**   The nature of the leaving group helps determine the ease with which S_N2 reactions occur. Iodine is the best leaving group and fluorine is the worst leaving group among the halogens. The best leaving groups tend to be conjugate bases of strong acids (or weak bases); in other words, they can effectively stabilize the electrons gained when they depart from the substrate. Because I is a better leaving group than Cl, the relative rate of this bimolecular reaction increases.

38. **D**   Addition of strong and sterically hindered nucleophiles increases the formation of elimination product(s). The nucleophile acts more as a base and abstracts a proton instead of attacking the carbon. It is easier to abstract a proton than to attack a carbon. Choice B leads to a decrease in the relative reaction rate because a good leaving group is substituted for a poor leaving group. Choice C decreases the relative rate for S_N2 because the carbon-bearing leaving group is more substituted and E1 occurs more often.

39. **B**   Methyl halides proceed via an S_N2 mechanism. Since only one carbon is present, elimination cannot occur. Methyl carbocations are the least stable; therefore, S_N1 can be ruled out.

40. **C**   Note that choice C is the conjugate base of acetic acid. Athough considered a weak acid, the conjugate base of acetic acid is not a very strong base and cannot abstract a proton β to the leaving group. All the other choices would enhance E2.

41. **D**   Adding stronger base or a sterically hindered base such as *tert*-butoxy ion favors elimination. Adding a catalyst does not change the outcome of the reaction and does not shift the directions of the reactions toward elmination. Lowering the temperature, however, tends to shift the reaction toward substitution.

Section I: General Concepts and Hydrocarbon Chemistry

Benzene

1. Which compound is known as aniline?

A.

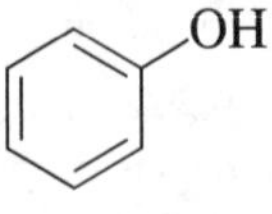

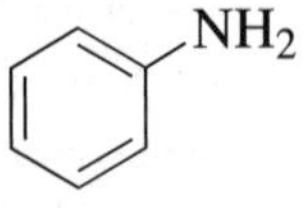

B.

C.

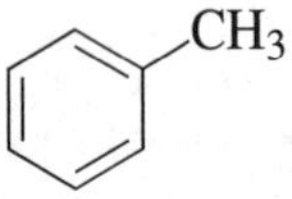

D.

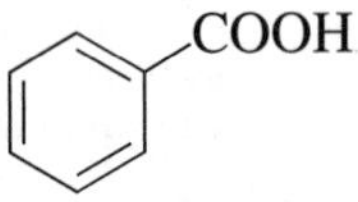

2. The **best** name for the following compound is:

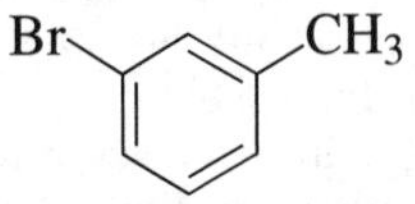

A. 3-methyl-bromobenzene.
B. 3-bromoaniline.
C. 3-methylbromobenzene.
D. 3-bromotoluene.

3. The correct IUPAC name for the following compound is:

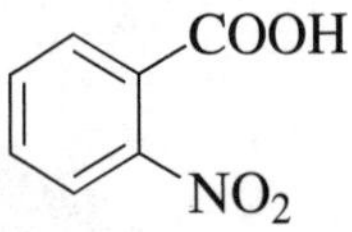

A. *p*-nitrobenzene methanoic acid.
B. *o*-nitrobenzoic acid.
C. *m*-nitrobenzoic acid.
D. *o*-nitrobenzene methanoic acid.

4. The correct IUPAC name for the following compound is:

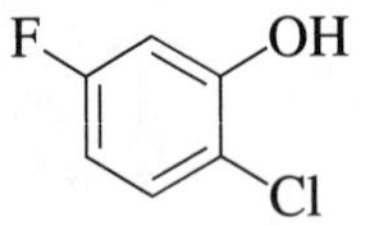

A. 2-chloro-5-fluoro-1-hydroxybenzene.
B. 1-hydroxy-2-chloro-5-fluorobenzene.
C. 1-fluoro-3-hydroxy-4-chlorobenzene.
D. none of the above.

5. The correct IUPAC name for the following compound is:

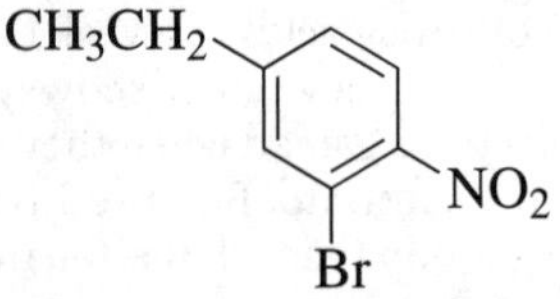

A. 3-bromo-4-ethyl-1-nitrobenzene.
B. 1-ethyl-2-bromo-4-nitrobenzene.
C. 2-bromo-4-nitrotoluene.
D. 3-bromo-1-ethyl-4-nitrobenzene.

6. The correct IUPAC name for the following compound is:

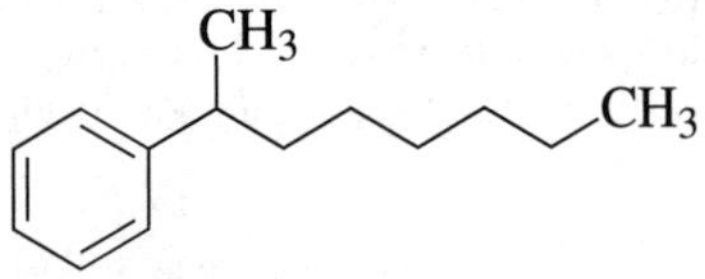

A. *sec*-octylbenzene.
B. 2-octylbenzene.
C. *iso*-octylbenzene.
D. 2-phenyloctane.

7. Which statement about the structure of benzene is NOT true?

A. The two Kekulé structures of benzene are in equilibrium.
B. The carbon–carbon bond lengths in benzene are greater than the carbon–carbon single bonds and less than the carbon–carbon double bonds in aliphatic compounds.
C. The molecular geometry of benzene is best described as planar.
D. The stability of benzene is much greater than the stability of 1,3,5-cycloheptatriene.

8. The molecular orbital structure of benzene includes:

   I. sp^2 hybridized carbon atoms with bond angles of 120°.
   II. six overlapping p-orbitals extending above and below the carbon ring.
   III. delocalized π-bond electrons.
   IV. 12 σ-bonds formed in each molecule of benzene.

   A. I, II, and III
   B. I, III, and IV
   C. II, III, and IV
   D. I, II, III, and IV

9. Which compound is aromatic?

   A.
   C.

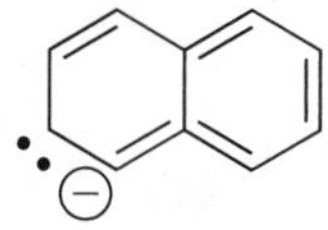

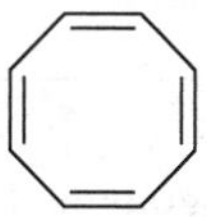

   B.
   D.

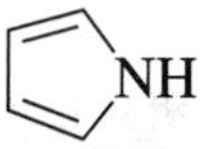

10. The phenomenon of delocalized electrons is NOT important in:

   A. resonance.
   B. the stabilization of resonance hybrids.
   C. the freedom of rotation about bonds.
   D. saturated aliphatic compounds.

11. Aromatic compounds, such as benzene, most often react by:

   A. nucleophilic aromatic substitution mechanisms.
   B. electrophilic aromatic addition mechanisms.
   C. nucleophilic aromatic addition mechanisms.
   D. electrophilic aromatic substitution mechanisms.

Questions 12–13 refer to the following information:

Compound X is electron-deficient and reacts with benzene in a two-step mechanism.

12. Which carbocation is NOT an intermediate that forms in the first step?

   A.
   C.

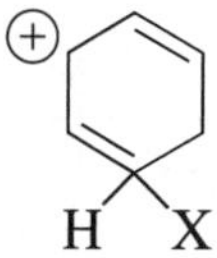

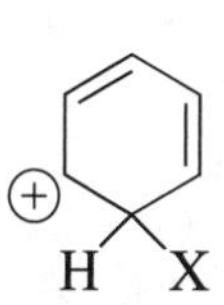

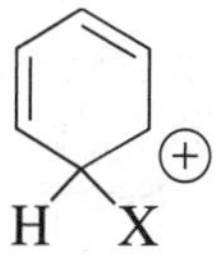

   B.
   D.

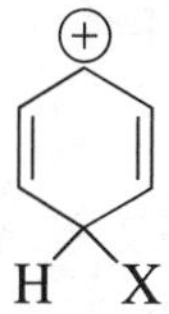

13. In the second step of the reaction described previously, which event occurs?

   A. The carbocation rearranges to a more stable intermediate.
   B. An intermediate resonance hybrid forms.
   C. A proton is lost from the carbon bearing the positive charge.
   D. A proton is lost from the carbon bearing compound X.

14. Which compounds do NOT react by electrophilic aromatic substitution mechanisms?

I. 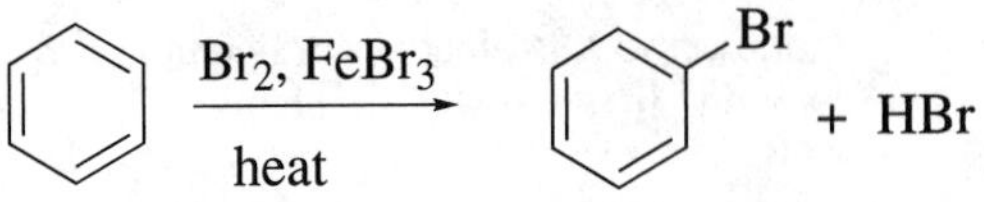 benzene $\xrightarrow[\text{heat}]{\text{Br}_2,\ \text{FeBr}_3}$ bromobenzene (Br) + HBr

II. 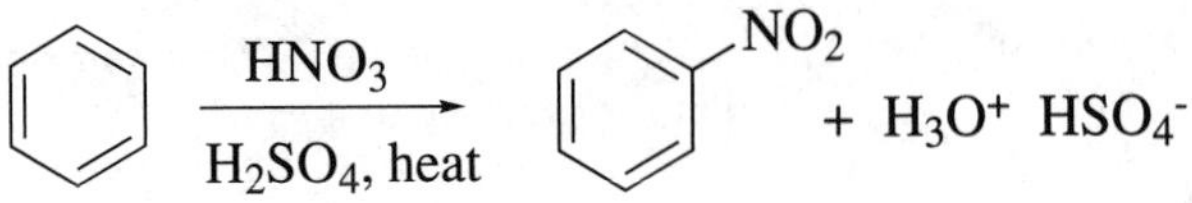 benzene $\xrightarrow[\text{H}_2\text{SO}_4,\ \text{heat}]{\text{HNO}_3}$ nitrobenzene (NO_2) + H_3O^+ HSO_4^-

III. 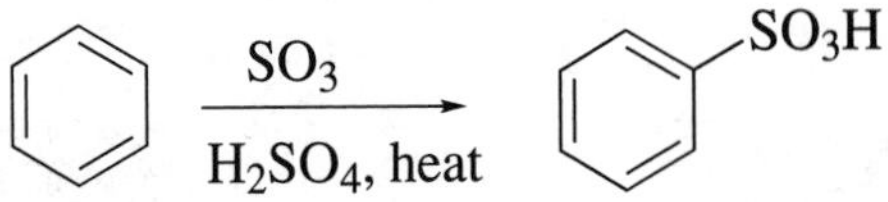 benzene $\xrightarrow[\text{H}_2\text{SO}_4,\ \text{heat}]{\text{SO}_3}$ benzenesulfonic acid (SO_3H)

IV. 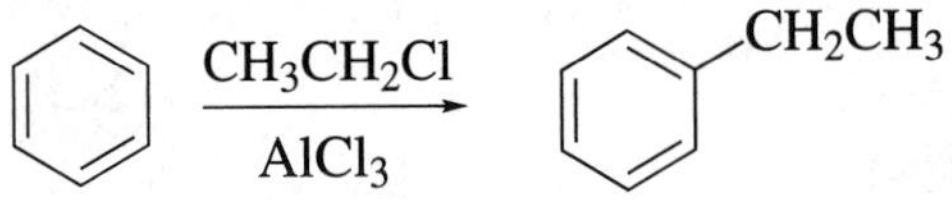 benzene $\xrightarrow[\text{AlCl}_3]{\text{CH}_3\text{CH}_2\text{Cl}}$ ethylbenzene (CH_2CH_3)

V. 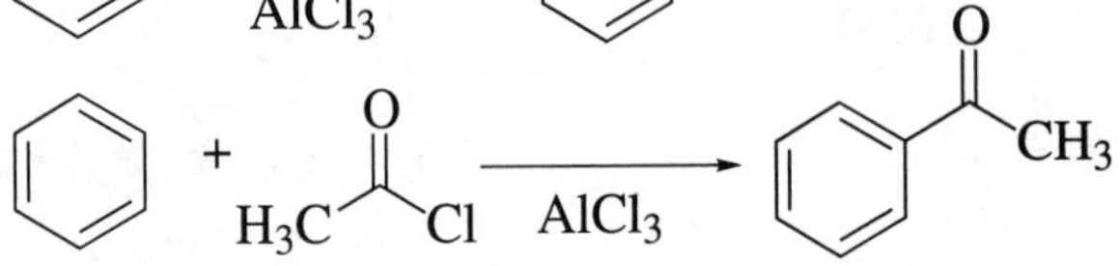 benzene + H_3C—C(=O)—Cl $\xrightarrow{\text{AlCl}_3}$ acetophenone (C(=O)CH_3)

A. I and V
B. I, III, and IV
C. II and IV
D. All are electrophilic substitution reactions.

15. Which compound undergoes substitution reactions faster than benzene?

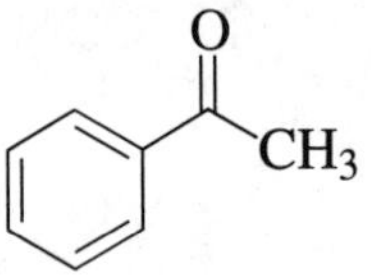

A. 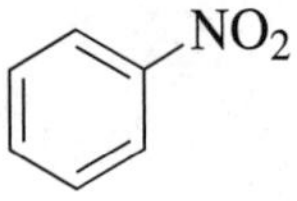 NO_2 (nitrobenzene)
B. 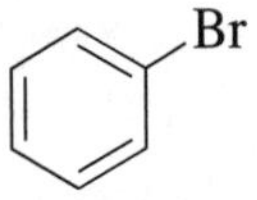 Br (bromobenzene)
C. C(=O)CH_3 (acetophenone)
D. 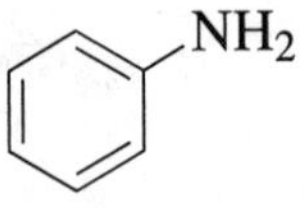 NH_2 (aniline)

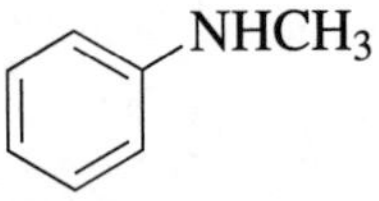

16. Which compound undergoes substitution reactions slower than benzene?

A. 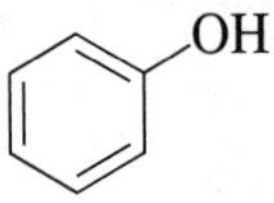 OH (phenol)
B. 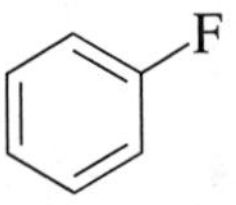 F (fluorobenzene)
C. $NHCH_3$
D. 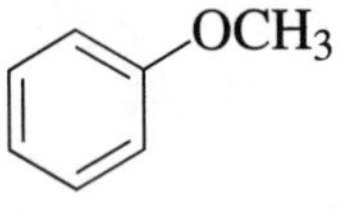 OCH_3

17. If a single bromine atom is substituted on the ring of the compound diagrammed in the following halogenation reaction, the bromine is most often found at position(s):

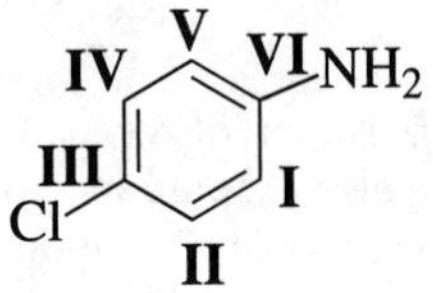

(ring with NH_2 at position I, Cl at position III; positions IV, V, VI, I, II marked)

A. II and IV   C. I and V
B. I and III   D. III

18. If a single chlorine atom is subsituted on the ring of the compound diagrammed in the following halogenation reaction, the chlorine is found at postion(s):

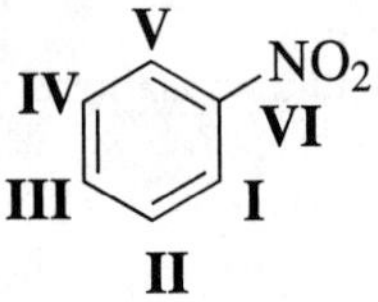

(ring with NO_2 at position VI; positions IV, V, VI, I, II, III marked)

A. I and V   C. III
B. I, III, and V   D. II and IV

19. If an *ortho–para* directing substituent is substituted on the following compound, at what position(s) is it found?

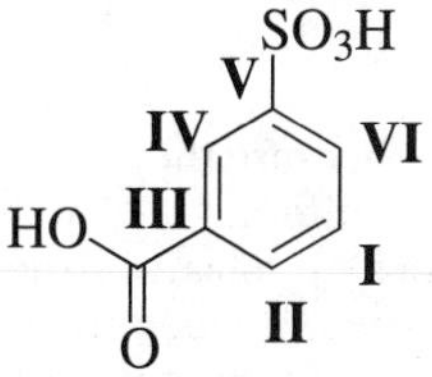

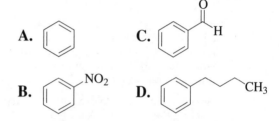

 A. II   **C.** II and VI
 B. VI   **D.** I

20. Why are electron-withdrawing groups *meta* directors?

 A. The carbocation intermediate has a negative charge on the *meta* position.
 B. The more stable resonance hybrid occurs with *meta* attachment of the electrophile.
 C. The less stable resonance hybrid occurs with *meta* attachment of the electrophile.
 D. The carbocation intermediate has a positive charge at the *meta* position.

21. Phenol is an *ortho–para* director because the hydroxy group:

 A. donates electrons that increase electron density at *ortho* and *para* positions favoring nucleophilic attack.
 B. donates electrons that increase electron density at *ortho* and *para* positions favoring electrophilic attack.
 C. donates electrons to the *ortho* and *para* positions and attracts electrons away from *meta* positions favoring nucleophilic attack of the ring.
 D. donates electrons to the *ortho* and *para* positions and attracts electrons away from *meta* positions favoring electrophilic attack of the ring.

22. Which acid is the strongest?

 A. 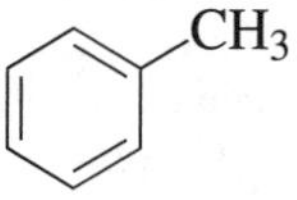 [benzene with CH_3]
 C. 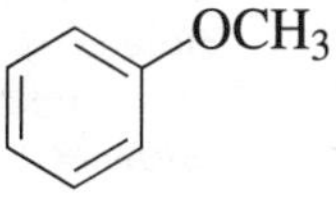 [benzene with OCH_3]
 B. 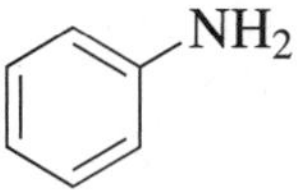 [benzene with NH_2]
 D. 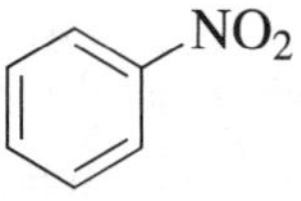 [benzene with NO_2]

23. Which base is the weakest?

 A. 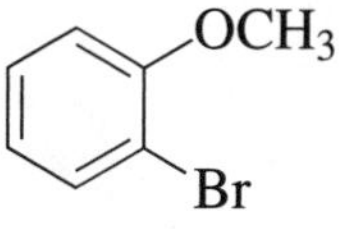 [benzene with OCH_3 and Br]
 C. 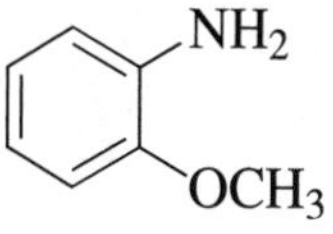 [benzene with NH_2 and OCH_3]
 B. 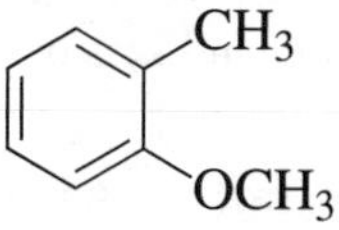 [benzene with CH_3 and OCH_3]
 D. 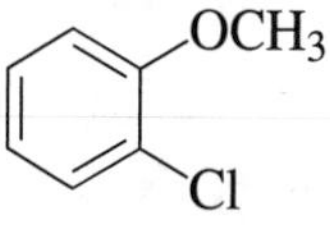 [benzene with OCH_3 and Cl]

24. Which base is the strongest?

 A. 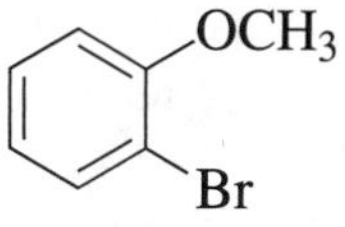 [benzene with OCH_3 and Br]
 C. 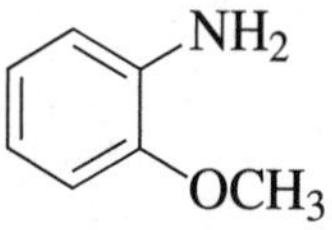 [benzene with NH_2 and OCH_3]
 B. 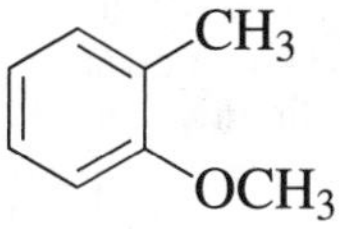 [benzene with CH_3 and OCH_3]
 D. 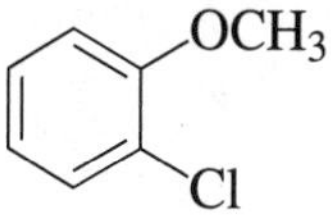 [benzene with OCH_3 and Cl]

25. Which acid is the weakest?

 A. [benzene]
 C. [benzene with CHO]
 B. [benzene with NO_2]
 D. [benzene with CH_3 chain]

SOLUTIONS

Benzene

1. **C**   Compound C is known as aniline. A is phenol, B is toulene, and D is benzoic acid.

2. **D**   When hydroxyl, methyl, or amino groups are attached to benzene, the compounds are given the names phenol, toluene, and aniline, respectively. In this compound, the methyl group on the ring makes it a toluene molecule. Begin numbering at the carbon bearing the methyl group and make the substituents have the lowest numbers possible. The result is that the bromine atom is at the 3 position. This molecule can also be designated *meta*-bromotoluene (Br is two positions away from the carbon bearing the methyl group).

3. **B**   When a carboxylic acid group is attached to benzene, the compound is known as benzoic acid. The nitro group is attached adjacent to or one carbon away from the acid group. The compound is therefore named *o*-nitrobenzoic acid or 2-nitrobenzoic acid.

4. **D**   This compound has a hydroxy group attached; therefore, it is considered a substituted phenol. Start numbering at the carbon bearing the hydroxy group and make the substituents have the lowest numbers possible. List them in alphabetic order. The name should be 2-chloro-5-fluorophenol, which is not one of the choices. Numbering in the other direction results in higher numbers on the substituents.

5. **D**   The numbering is done to produce the smallest total. Therefore, number from the ethyl group toward the nitro group. List the substituents in alphabetic order.

6. **D**   When six or more carbons are in a base chain, benzene is considered a substituent and is called a phenyl group. The longest chain is an 8-carbon chain; therefore, the base name is octane and the phenyl substituent is at position 2.

7. **A**   Neither of the two Kekulé structures for benzene has even been isolated. No equilibrium exists between these two structures because they are resonance structures, differing only in the positions of electrons. Resonance structures are not in equilibrium, because they do not represent two distinct molecules. The carbon–carbon bond lengths in benzene are 1.39 Å whereas the carbon–carbon single-bond length is 1.54 Å and the carbon–carbon double bond is 1.34 Å. Benzene is more stable than 1,3,5-cycloheptatriene because it possesses resonance or stabilization energy and 1,3,5-cycloheptatriene does not (not aromatic).

8. **D**   All choices are correct. Three sp^2 orbitals of each carbon atom in benzene from σ-bonds, all lying in the plane formed by the six carbon nuclei. Therefore, a σ-bond is formed between each of the 6 carbon atoms and between each carbon and hydrogen atom, making a total of 12 σ-bonds in benzene.

9. **D**   The Hückel rule states that aromatic compounds have a total of $4n+2\pi$-electrons (where n = integer) in a cyclic compound where each atom of the ring is associated with at least one sp^2 hybridized atom. Each double bond contains two π electrons, and the free lone pair of electrons on N is also counted. Only choice D has the correct number of electrons.

10. **D**   Electrons that cannot be assigned to one specific bond between two specific atoms are called delocalized electrons. The phenomenon of delocalized electrons is known as resonance. Delocalization of electrons or of charge stabilizes molecules or ions. The resonance structures produced give rise to a resonance hybrid with a rotation restriction about its bonds. Saturated aliphatic compounds tend not to have delocalized electrons.

11. **D**   Substitution reactions allow the aromatic π electrons to be regenerated after electrophilic attack of the ring. Following is the electrophilic aromatic substitution reaction for benzene. The first step is

electrophilic attack, which results in a resonance stabilized carbocation intermediate (resonance hybrid). The proton attached to the carbon with the new substituent is removed, and the electrons are used to regenerate the energetically favorable aromatic condition.

Step 1

Step 2
loss of proton

base

+ H

12. **A** The preceding mechanism shows the intermediate resonance forms. In addition, structure A has five bonds to the carbon bearing the H and X.

13. **D** See the solution for question 11.

14. **D** All five represent electrophilic aromatic substitution reactions with benzene. This reaction is characteristic of benzene. Reaction I is a halogenation, II is a nitration, III is a sulfonation, IV is a Friedel Crafts alkylation, and V is a Friedel Crafts acylation.

15. **D** Activating (electron-donating) groups on benzene result in enhanced electrophilic substitution. When deactivating (electron-withdrawing) groups are bound to benzene, the substitution reactions occur more slowly than they do for benzene. Of the choices shown, the amine is the only activating group. Although N is more electronegative than C, the nitrogen is capable of donating electrons to the ring via resonance.

16. **B** The halogen is the only deactivating group because of its electronegativity and its low resonance potential.

17. **C** The amino group is an *o/p* directing activator and the Cl is an *o/p* directing deactivator. The stronger activating group takes priority. Because the Cl is located *para* to the amino group, the only positions that Br can fill are those *ortho* to the amino group (strong activator), that is, positions I and V.

18. **D** Nitro groups are *meta* directors, and the positions *meta* to the nitro group are II and IV.

19. **D** The directing preference of the incoming substituent has little to do with the position that it takes on the ring. Both substituents that are already on the ring are *meta* directors, and position I is *meta* to both *meta-directing* substituents already on the ring.

20. **B** When the electrophile attaches to the benzene ring, a carbocation is formed. This carbocation is relatively stable because the charge can be delocalized about the ring. *Ortho* or *para* attack results in a resonance form, with the positive charge on the carbon bearing the electron-withdrawing group (highly unstable). This resonance form does not occur with *meta* placement of the incoming electrophile.

21. **B** The lone pair of electrons can donate, via resonance, to stabilize the carbocation intermediate. This stabilization occurs when the substituent accepts a positive by donating electrons that result in further delocalization of the positive charge. This situation does not occur with *meta* attack.

22. **D** The strongest acid possesses the strongest electron-withdrawing group because it can most readily stabilize the conjugate base. In this question, choice D is the only electron-withdrawing group.

23. **D** The weakest base is the strongest acid. The nitro group is the strongest deactivator and withdrawer of electrons. The halogens are the only other deactivating groups and they are weak deactivators at best. Chlorine is more electron-withdrawing than bromine because it is more electronegative. Therefore, the weakest base has the two strongest electron-withdrawing groups. The other choices in the order of decreasing basicity or increasing acidity are A, B, and C.

24. **C** The amino group is the best electron-donating group and gives the strongest base, in combination with the next strongest electron-donating group given here: the methoxy group. The other choices are either deactivators or weaker activators than the amino or methoxy.

25. **D** Alkyl groups are activators and, in this question, the only activator given. Because the weakest acid is the strongest base, choice D is the best answer. Choice C would be the most acidic.

Section II: The Chemistry of Oxygen-Containing Organic Compounds

Alcohols

1. A correct name for the following compound is:

OH
H₃C — CH — CH₂ — CH₃

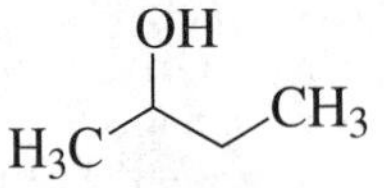

A. *iso*-butyl alcohol.
B. *sec*-butyl alcohol.
C. 2-butanol.
D. B and C.

2. A correct name for the following compound is:

H₂C=CH—CH₂—CH—CH₂—CH₃
         |
         OH

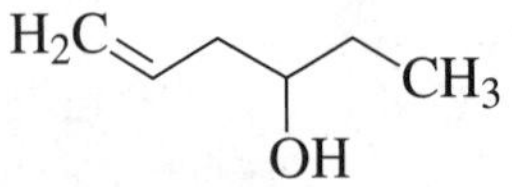

A. 1-hexen-4-ol.
B. 3-hydroxy-5-hexene.
C. 3-hexenol.
D. 5-hexen-3-ol.

3. A correct name for the following compound is:

C₆H₅—CH₂—OH

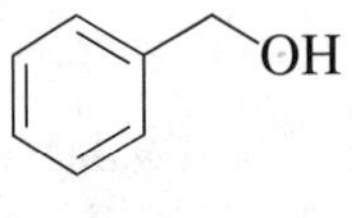

A. phenol.
B. 2-phenylmethanol.
C. 1-methanol benzene.
D. benzyl alcohol.

4. A correct name for the following compound is:

HO-CH₂CH₂CH₂CH₂-OH

A. 1,4-dihydroxybutane.
B. 1,4-butanediol.
C. butane-1,4-diol.
D. 1,4-dibutanol.

5. Why do alcohols have boiling points much higher than hydrocarbons of similar molecular weight?

A. Alcohols have greater van der Waals attraction forces.
B. Alcohol molecules have greater molecular symmetry.
C. Hydrogen bonds must be broken in the process of volatilization.
D. Alcohols must overcome greater ionic forces in the process of volatilization.

6. Why does decyl alcohol have a greater boiling point than pentyl alcohol?

A. Decyl alcohol has greater van der Waals attraction forces.
B. Decyl alcohol has greater hydrogen bonding forces to overcome.
C. Both of the above.
D. Neither of the above.

---

7. Rank the following alcohols by decreasing boiling point:

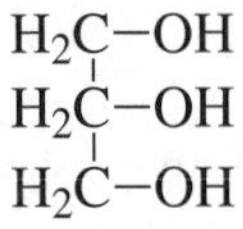

H₂C—OH
|
H₂C—OH   H₃C   OH   H₂C—OH
| |
H₂C—OH   H₂C—OH
                                 |
                                 CH₃

1 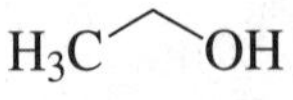   **2** 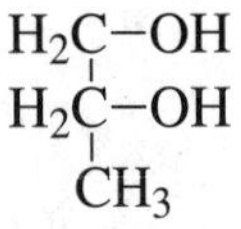   **3**

$CH_3(CH_2)_2OH$

OH   **4**
|
H₃C   CH₃

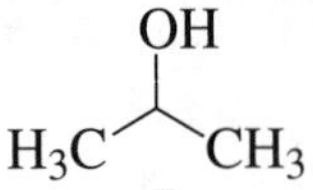

5

A. 1, 3, 5, 4, 2
B. 4, 5, 2, 3, 1
C. 1, 3, 4, 5, 2
D. 1, 3, 4, 2, 5

8. Which alcohol is the LEAST soluble in water?

A. H_2C-CH_2
     $\quad\ |\quad\ |$
     $\quad HO\ \ OH$

B. $H_3C\diagup\diagdown OH$

C.
$$H_3C\diagdown\underset{}{\overset{OH}{\diagup}}\diagdown CH_3$$

D. $HO\diagup\diagdown\diagup\diagdown CH_3$

9. Rank the following in decreasing order of acidity:

H_2O	CH_3CH_3	CH_3CH_2OH	$CH_2=CH_2$
1	**2**	**3**	**4**

NH_3	$HC\equiv CH$
5	**6**

A. 1, 3, 6, 5, 4, 2
B. 6, 1, 3, 5, 4, 2
C. 2, 4, 5, 3, 1, 6
D. 2, 4, 5, 6, 3, 1

10. Rank the following in increasing order of acidity:

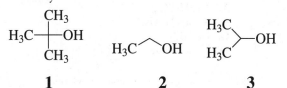

$$\underset{\textbf{1}}{H_3C-\overset{CH_3}{\underset{CH_3}{\overset{|}{\underset{|}{C}}}}-OH} \qquad \underset{\textbf{2}}{H_3C\diagup\diagdown OH} \qquad \underset{\textbf{3}}{\overset{H_3C}{\underset{H_3C}{>}}-OH} \qquad \underset{\textbf{4}}{H_2O}$$

A. 4, 2, 3, 1
B. 1, 2, 3, 4
C. 4, 1, 3, 2
D. 1, 3, 2, 4

11. Which organic compound has the greatest melting point?

A. Methanol
B. Ethanol
C. Propanol
D. All have equal melting points.

Questions 12–13 refer to the following statement:

Alcohols can be prepared from the hydration of alkenes in the presence of both water and an acid catalyst.

12. What happens in the first step of the mechanism for this reaction?

A. A carbocation forms by the addition of a water molecule.
B. A carbocation forms by the addition of a proton.
C. A carbanion forms by the loss of a proton from the alkene.
D. A carbanion forms by gaining electrons from water.

13. Which molecule is added to the intermediate generated in the first step of this reaction?

A. OH^-
B. H^+
C. H_2O
D. H_3O^+

14. The acid-catalyzed dehydration mechanism for alcohols is BEST described as a(n):

A. E1.
B. E2.
C. S_N1.
D. S_N2.

Questions 15–19 refer to the following three-step mechanism for alcohol dehydration:

1. $H_3C\diagup\overset{H\ \ H}{\diagdown}OH\ +\ H_3O^+\ \rightleftharpoons\ X$

2. $X\ \rightleftharpoons\ Y\ +\ H_2O$

3. $Y\ +\ H_2O\ \rightleftharpoons\ \overset{H}{\underset{H}{>}}=\overset{H}{\underset{H}{<}}\ +\ H_3O^+$

15. The **X** represents a(n):

A. carbocation intermediate.
B. carbanion intermediate.
C. protonated alcohol.
D. alcohol.

16. Which step is the slowest?

A. 1
B. 2
C. 3
D. All steps occur at the same rate.

17. The third step must involve:

    A. the loss of a proton by an alcohol.
    B. the gain of a proton by an alcohol.
    C. the loss of a proton by a protonated alcohol.
    D. the loss of a proton by a carbocation.

18. Step **2** is considered:

    A. endothermic.
    B. exothermic.
    C. adiabatic.
    D. none of the above.

19. Which molecule is expected to have higher reactivity than the alcohol shown in the preceding three-step mechanism for alcohol dehydration?

    A. 1° alcohol
    B. 2° alcohol
    C. 3° alcohol
    D. Both 2° and 3° alcohols

20. Which of the following oxidizing agents does NOT convert 1° alcohols exclusively to aldehydes?

    I. $K_2Cr_2O_7$, H_2SO_4
    II. $KMnO_4$, OH^-, H_2O, heat
    III. Cu, 300°C
    IV. CrO_3, pyridine

    A. III
    B. II
    C. IV
    D. II and III

21. Which oxidizing agent is commonly used to oxidize secondary alcohols?

    A. Ag, H_2SO_4
    B. $K_2Cr_3O_6$, HCl
    C. H_2CrO_4
    D. None of the above

22. The oxidation of a secondary alcohol produces a(n):

    A. carboxylic acid.
    B. aldehyde.
    C. ketone.
    D. ester.

SOLUTIONS

Alcohols

1. **D** Note that the hydroxy group is attached to the second carbon in the chain. Therefore, the butane can be considered a *sec*-butyl group. A common name for this compound is *sec*-butyl alcohol. The IUPAC name is 2-butanol. The longest chain is four-carbons, and the "e" suffix is dropped for "ol."

2. **D** Start numbering from the end of the longest carbon chain closest to the hydroxy group. The hydroxy group takes priority over the double bond. Therefore, the compound has its double bond in the 5 position and the hydroxy group in the 3 position.

3. **D** Benzyl alcohol is the common name for this compound (similar to benzyl carbocation or benzyl radical).

4. **B** Two hydroxy groups are attached to the one and four carbons. Compounds with two hydroxy groups are known as *diols*. The positions of the alcohol groups in simple diols are listed before the hydrocarbon base name.

5. **C** Alcohols form strong hydrogen bonds (H-bonds) with one another. To boil a liquid, the vapor pressure must be raised to atmospheric pressure. To accomplish this increase, the liquid must be volatilized, that is, the hydrogen bonds must be broken. Although H-bonds are generally considered relatively weak, a large amount of them connect alcohol molecules, added up, they represent an attractive force that results in higher boiling points.

6. **A** Both of these alcohols are primary alcohols with one -OH group. Their H-bonding capacity is similar. What differs between these compounds is the length of the carbon chain. As the length of the chain increases, so do the van der Waals attractive forces between alcohol molecules.

7. **C** The alcohol with the greatest boiling point has either a much higher molecular weight than the other alcohols or has greater H-bonding ability. Glycerol (**1**)

has three hydroxy groups and therefore has the greatest H-bonding capacity. 1,2-propanediol (**3**) has two hydroxy groups and therefore has slightly less H-bonding capacity. Choices **4** and **5** both have three carbons and one hydroxy group. Branched alcohols, however, have decreased capacity for both H-bonds and van der Waals interactions (due to sterics) and therefore have lower boiling points. The lowest is ethanol (**2**) because of fewer van der Waals interactions and only one hydroxy group.

8. **D** The alcohols with more than four carbons in the hydrocarbon chain have little to no solubility in water. Choice A, a diol, has the greatest solubility because of the small number of carbons and its ability to associate with water (presence of two hydroxy groups).

9. **A** $H_2O > ROH >$ alkynes $>$ ammonia $>$ alkenes $>$ alkanes. Think about the stability of the conjugate bases when trying to assess differences in acidity. Alkyl groups are electron-donating; therefore, alkoxides are more reactive than hydroxides. Alkynes and ammonia are even weaker acids because the conjugate bases are unstable. In addition, the conjugate bases are very strong. Moreover, greater s-character allows the carbon to hold on tighter to the electron gained after deprotonation; therefore, alkynes are more acidic than alkenes, which are more acidic than alkanes.

10. **D** As the number of alkyl groups attached to the carbon bearing the hydroxy group increases, the acidity decreases. Alkyl groups are electron-releasing and they tend to stabilize the alcohol, making it more difficult to lose a proton. In addition, the presence of more alkyl groups destabilizes the negatively charged conjugate base.

11. **A** The melting point of most organic compounds increases with increasing molecular weight. This increase is true for the 1° alcohols, starting with butanol. Methanol has a melting point of $-98°C$,

ethanol (mp = $-117°C$), and propanol (mp = $-129°C$). This decreasing melting point with increasing molecular weight is an exception to the rule and reverses itself, beginning with butanol.

12. **B**   Remember, from the earlier section on alkenes, that H^+ can attack a carbon of the double bond, forming a carbocation that occurs following Markovnikov's rule. A water molecule then attacks the carbocation intermediate, leading to the alcohol after deprotonation.

13. **C**   Diagrammed in the answer for question 12

14. **A**   The acid-catalyzed dehydration reaction proceeds through a carbocation intermediate. The alcohol is first protonated and then acts as a leaving group. Before protonation, the hydroxy group is a poor leaving group.

15. **C**   As the mechanism demonstrates, a protonated alcohol is formed.

16. **B**   The formation of the carbocation is the slowest step and therefore rate-determining.

17. **D**   Again, the mechanism given in the solution to question 15 shows the loss of a proton by a carbocation that occurs in the third step.

18. **A**   The reactions by which carbocations are formed from protonated alcohols are highly endothermic. This concept could have been postulated from the fact that step 2 is the rate-limiting step.

19. **D**   This reaction requires a carbocation intermediate. The 3° carbocations are more stable than 2° carbocations, which are, in turn, more stable than 1° carbocations. The alcohol in this reaction is 1°; therefore, a 3° or 2° alcohol can be expected to yield more stable intermediates. This stability decreases

the energy of activation of the rate-limiting step (greater reactivity).

20. **B**   Choice A oxidizes 1° alcohols to aldehydes and further oxidizes the aldehydes to carboxylic acids if they are not removed. Choice B is a very strong oxidizing agent that directly oxidizes alcohols to carboxylic acids. The other two stop oxidizing at the aldehyde.

21. **C**   Choices A and B are not oxidizing agents. Chromium is most commonly in a +6 oxidation state before oxidation of organic compounds. For B, the oxidation state of Cr is not +6. C allows 2° alcohols to be oxidized to ketones.

22. **C**   The structure of a 2° alcohol allows only ketones to be regularly produced via oxidation. Otherwise, a carbon–carbon bond has to be broken. It is very energy-expensive to break carbon–carbon bonds.

144

Section II: The Chemistry of Oxygen-Containing Organic Compounds

Aldehydes and Ketones

1. The correct IUPAC name for the following compound is:

A. 2-chloro-4-butanal.
B. 2-chlorobutan-4-al.
C. 3-chlorobutanal.
D. 3-chlorobutan-1-al.

2. The correct IUPAC name for the following compound is:

A. benzaldehyde.
B. benzanal.
C. 1-phenylmethanal.
D. phenylaldehyde.

3. The correct IUPAC name for the following compound is:

A. 2-butanone.
B. 3-butanone.
C. butanone.
D. 1-methylpropanone.

4. The correct IUPAC name for the following compound is:

A. ethyl propyl ketone.
B. 3-hexanone.
C. 4-hexanone.
D. hexanone.

5. The correct common name for the following compound is:

A. ethyl propyl ketone.
B. 2-pentanone.
C. butyl methyl ketone.
D. methyl propyl ketone.

6. Rank the following molecules in order of decreasing boiling point:

$CH_3CH_2CH_3$ $CH_3CH_2CH_2OH$ 3
     1   2 CH_3CH_2 (O) H

A. 3, 2, 1   C. 1, 3, 2
B. 2, 1, 3   D. 2, 3, 1

7. What property of low–molecular weight aldehydes and ketones accounts for the magnitude of their boiling points?

A. The ability to form strong H-bonds between their molecules
B. The ability of the carbonyl oxygen to form H-bonds with other carbonyl groups
C. The ability of the polar carbonyl group to attract other polar molecules
D. The ability of the carbonyl group to attract electrophiles and form bonds

8. Which compound has the greatest solubility in water?

9. Which action best accounts for the solubility of aldehydes and ketones in water?

   A. Polar interactions between solute molecules
   B. H-bonding between solute molecules
   C. Van der Waals forces
   D. H-bonding between solute and solvent molecules

10. The melting points of aldehydes and ketones tend to:

    A. decrease with increasing molecular weight.
    B. increase with increasing molecular weight.
    C. remain unchanged with increasing molecular weight.
    D. be unpredictable due to tautomer resonance.

11. Which statement about the carbonyl group is NOT true?

    A. The carbonyl carbon is sp^2 hybridized.
    B. The bond angles among the three atoms attached to the carbonyl carbon are 120°.
    C. The three atoms attached to the carbonyl form a nonplanar geometry.
    D. The carbonyl group forms resonance structures.

12. Which statement about the carbonyl group of ketones and aldehydes is true?

    I.   It can attract nucleophiles.
    II. It can attract electrophiles.
    III. It tends to undergo addition reactions.
    IV. It tends to undergo substitution reactions.

    A. I and III
    B. II and IV
    C. I, II, and III
    D. I, III, and IV

Questions 13–18 refer to the following reactions. X and Y are unknown reaction products, not intermediates.

1. 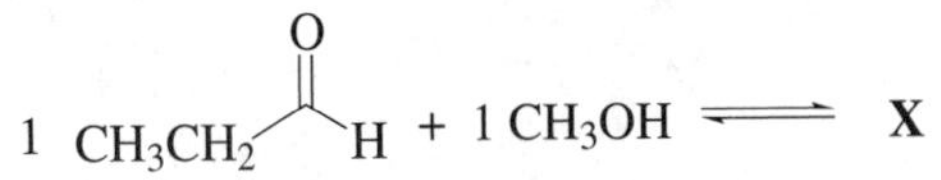 1 CH$_3$CH$_2$—CHO + 1 CH$_3$OH $\rightleftharpoons$ X

2.   1 X + 1 CH$_3$OH $\rightleftharpoons$ Y

13. In the reaction of propanal and methanol, the alcohol acts as a(n):

    A. proton donor.
    B. electron acceptor.
    C. electrophile.
    D. nucleophile.

14. Reaction 1 is considered a(n):

    A. nucleophilic substitution.
    B. electrophilic addition.
    C. nucleophilic addition.
    D. electrophilic substitution.

15. Compound X is called a(n):

    A. protonated alcohol.
    B. acetal.
    C. ketal.
    D. hemiacetal.

16. Product Y is called a(n):

    A. protonated alcohol.
    B. acetal.
    C. hemiketal.
    D. hemiacetal.

17. If propanal is replaced by propanone in the preceding reactions, the basic mechanism:

    A. changes.
    B. stays the same although the products differ.
    C. cannot be predicted.
    D. shows no reaction.

18. If propanone replaces propanal in reaction 1, product Y is a(n):

    A. ketal.
    B. acetal.
    C. hemiketal.
    D. hemiacetal.

Question 19–21 refer to the following reactions and data:

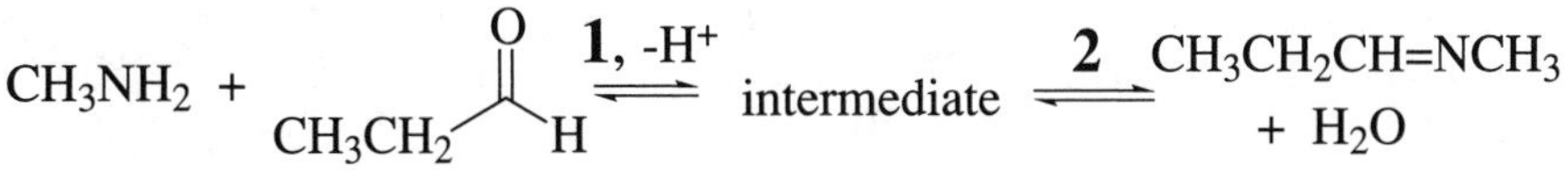

$$CH_3NH_2 + \quad \underset{CH_3CH_2}{\overset{O}{\|}} \quad \overset{\textbf{1, -H}^+}{\rightleftharpoons} \quad \text{intermediate} \quad \overset{\textbf{2}}{\rightleftharpoons} \quad \underset{+ \; H_2O}{CH_3CH_2CH=NCH_3}$$

Primary amines or ammonia react with aldehydes and ketones to form Schiff bases (imines).

19. Reaction 1 is BEST considered:

    A. a nucleophilic substitution reaction.
    B. an electrophilic addition reaction.
    C. a nucleophilic addition reaction.
    D. an electrophilic substitution reaction.

20. Reaction 2 is BEST considered:

    A. a nucleophilic substitution reaction.
    B. an electrophilic addition reaction.
    C. an elimination reaction.
    D. an electrophilic substitution reaction.

21. The BEST structure for the intermediate is:

A. 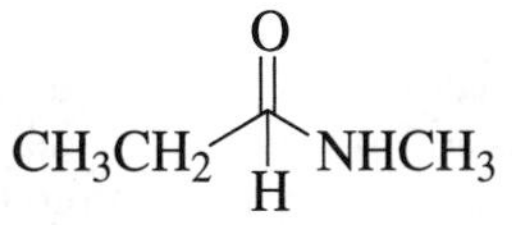 $CH_3CH_2\underset{H}{\overset{O}{\|}}NHCH_3$ **C.** 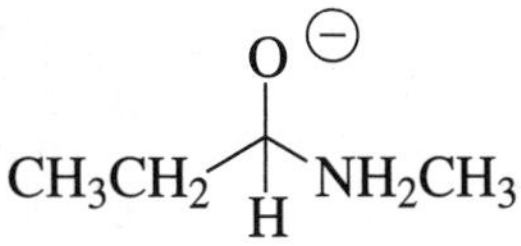 $CH_3CH_2\underset{H}{\overset{O^{\ominus}}{|}}NH_2CH_3$

B. 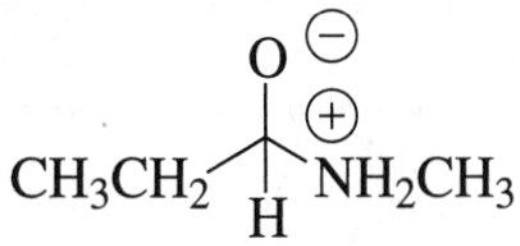 $CH_3CH_2\underset{H}{\overset{O^{\ominus}}{|}}\overset{\oplus}{N}H_2CH_3$ **D.** 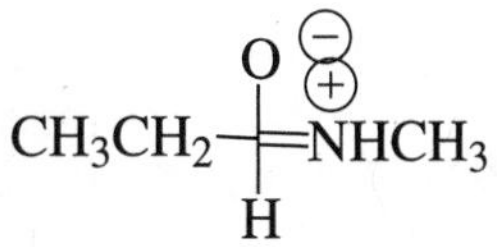 $CH_3CH_2\underset{H}{\overset{O^{\ominus}\overset{\oplus}{}}{=}}NHCH_3$

22. Which hydrogen is the most easily abstracted by a base?

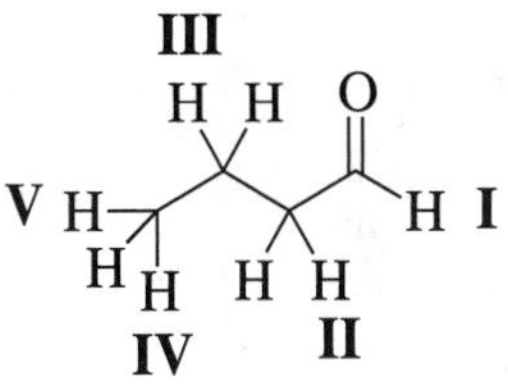

    A. I
    B. II
    C. III
    D. IV or V

23. In the presence of trace acids and bases, ketones;

    A. form hemiketals.
    B. form hemiketals and ketals.
    C. undergo keto–enol tautomerism.
    D. undergo nucleophilic addition reactions.

24. The resonance-stabilized anion shows the reason for:

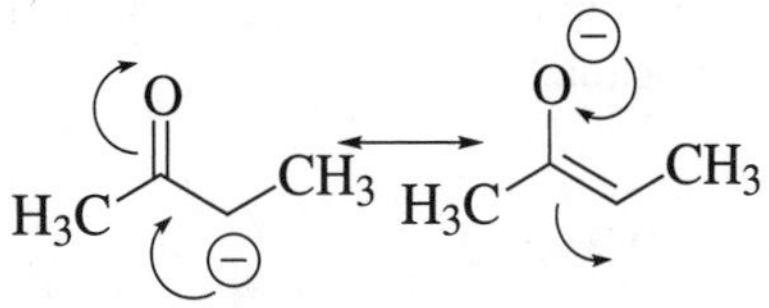

    A. the unusual stability of ketones.
    B. the unusual stability of enols.
    C. the basicity of protons on the α-carbon.
    D. The acidity of protons on the α-carbon.

25. Which form is the most stable?

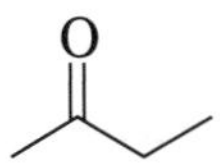

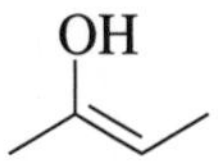

    A. I
    B. II
    C. They are equally stable.
    D. Both are very unstable.

Questions 26–31 refer to the following reaction:

A base is slowly added to a solution of acetaldehyde and benzaldehyde. The base first reacts with acetaldehyde; then compound **A** reacts with benzaldehyde to

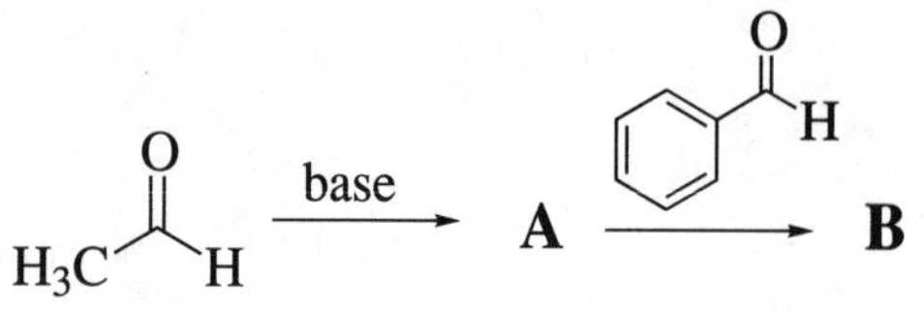

give compound **B**.

26. In the preceding reaction, the base acts to:

   A. deprotonate the carbonyl group.
   B. abstract a proton from the α-carbon.
   C. attack the carbonyl group of the aldehyde.
   D. initiate nucleophilic substitution.

27. In the second step, compound **A** acts as a(n):

   A. carbonium ion.
   B. nucleophile.
   C. electrophile.
   D. free radical.

28. In the second step, compound **A**:

   A. attacks the aromatic ring in an electrophilic aromatic addition.
   B. attacks the α-carbon of the aldehyde in a nucleophilic addition reaction.
   C. attacks the α-hydrogen of the aldehyde in a nucleophilic substitution reaction.
   D. attacks the carbonyl carbon of benzaldehyde in a nucleophilic addition reaction.

29. What is the structure of product **B**?

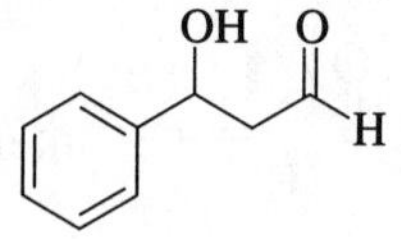

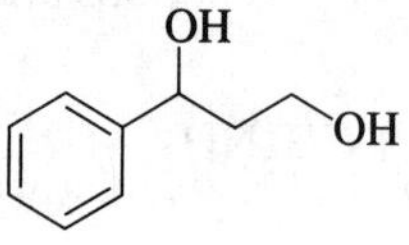

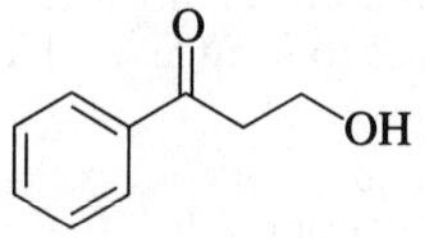

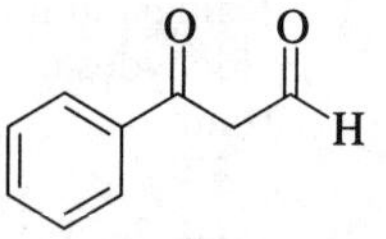

30. If product **B** is added to a warm basic solution, the new product is most likely:

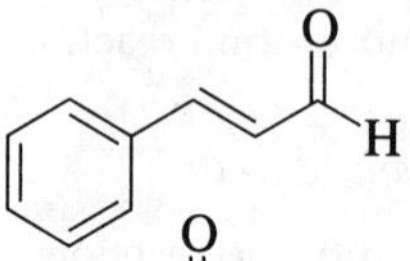

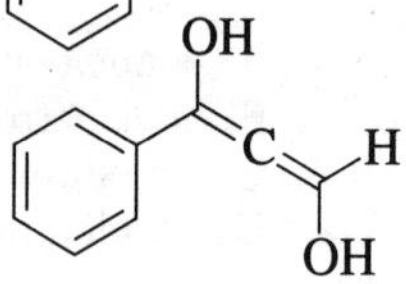

31. Why is the base slowly added to a solution of both of the aldehydes?

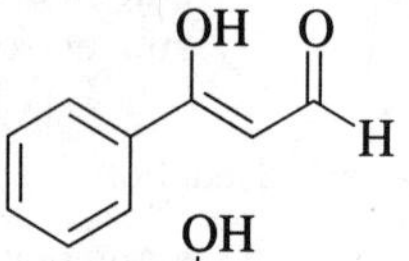

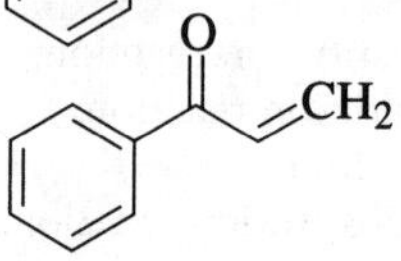

   A. Self-condensation of acetaldehyde occurs if the base is added to just acetaldehyde.
   B. Benzaldehyde contains acidic protons.
   C. Self-condensation of benzaldehyde occurs if the base is added to just benzaldehyde.
   D. There is no apparent reason for this technique.

32. Which form is the most difficult to oxidize?

   A. Propanol
   B. Propanal
   C. 2-Propanone
   D. All are equally difficult to oxidize.

SOLUTIONS

Aldehydes and Ketones

1. **C** Choose the longest carbon chain to which the carbonyl group is attached. Remember to count the carbonyl carbon as carbon 1. Cl is attached to C-3; and the "e" from the base name is replaced with "al."

2. **A** Dropping the "e" and replacing with "al" is appropriate only for aliphatic aldehydes. "Benz" is used for the prefix instead of "phenyl" for ketones, carboxylic acids; and derivatives.

3. **C** This question is rather picky. For ketones of three or four carbons, the carbonyl must be on the second or third position. Therefore, naming a compound 2-butanone or 3-butanone is redundant and undesirable (according to IUPAC).

4. **B** Choose the longest carbon chain to which the carbonyl group is attached, and begin numbering at the end closest to the carbonyl group. The carbonyl carbon position is 3.

5. **D** To identify ketones by their common name, identify the two alkyl groups directly attached to the carbonyl. List in alphabetical order.

6. **D** Aldehydes and ketones can H-bond with water molecules because they contain the polar carbonyl group. However, because they do not contain a hydrogen directly attached to the polar atom (oxygen), they do not form H-bonds with themselves. Alcohols have the ability to H-bond to themselves; therefore, they have higher boiling points than aldehydes or ketones of similar molecular weight. Both aldehydes and ketones are more polar than alkanes and can H-bond to water; for this reason, they have higher boiling points than hydrocarbons of similar molecular weight.

7. **C** Boiling point is affected by intermolecular forces (e.g., H-bonds) and molecular weight. Choices A and B are false, and choice D does not explain the boiling points of aldehydes and ketones. Choice C is therefore the best answer, as described in the solution to question 6.

8. **B** Choice A has a hydrophobic aromatic ring as a substituent; therefore, it is insoluble in water. Of the other choices, the most water-soluble tend to be those compounds with the shortest carbon chains. Choices B and D both have four carbons, whereas choices C and D have five. Between the ketone and aldehyde in choices B and D, the ketone is more soluble in water. It is generally true that the small ketones are more water-soluble than the small aldehydes of similar carbon length.

9. **D** H-bonding can occur between the aldehyde or ketone (solute) molecule and the water (solvent) molecule. This interaction between solute and solvent gives a compound water-solubility.

10. **B** The general trend of melting points is to increase with increasing molecular weight.

11. **C** Only choice C is false. The geometry of the carbonyl group and its substituents is trigonal planar (sp^2 hybridization, and the angle between the substituents is 120°.

12. **C** The carbonyl oxygen has a partial negative charge as it drags electrons from the carbonyl carbon toward itself. This action leaves the carbon with a partial positive charge. Electron-deficient molecules can attack the carbonyl oxygen whereas electron-rich molecules attack the carbonyl carbon. This difference points out that nucleophiles attack the carbonyl via nucleophilic addition reactions. Electrophiles such as Lewis acids complex with the lone pairs on the carbonyl oxygen.

13. **D** A characteristic reaction of aldehydes and ketones is nucleophilic addition. The nucleophile (electron-rich) attacks the electron-poor carbonyl carbon to give a tetrahedral intermediate as the hybridization of the carbon changes from sp2 to sp3. After proton transfer, the

result is the *hemiacetal.* Therefore, in this example, the methanol acts as a nucleophile.

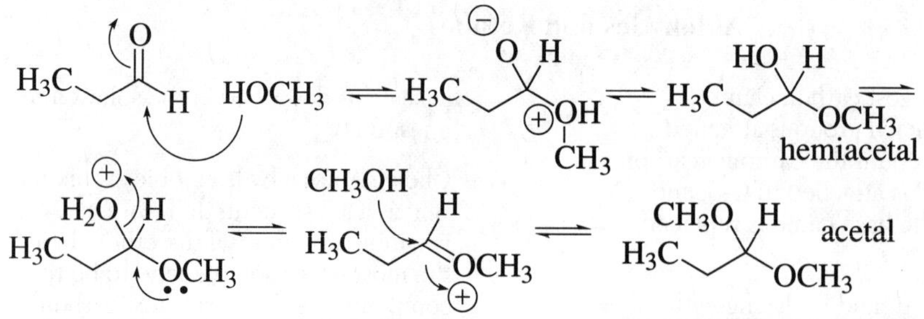

H_3C—C(=O)—H HOCH$_3$ $\rightleftharpoons$ H_3C—C(O$^-$)(H)—$^+$OH—CH$_3$ $\rightleftharpoons$ H_3C—C(HO)(H)—OCH$_3$ hemiacetal

H_3C—C($^+$H$_2$O)(H)—OCH$_3$ $\rightleftharpoons$ H_3C—C(H)($^+$)—OCH$_3$ (CH$_3$OH) $\rightleftharpoons$ H_3C—C(CH$_3$O)(H)—OCH$_3$ acetal

14. **C** As shown in the mechanism for question 13, the reaction is nucleophilic addition of methanol to the carbonyl carbon.

15. **D** The intermediate X is known as the hemiacetal. A full acetal is the result of a second nucleophilic attack of the carbon, which was the carbon of the carbonyl. Hemiketals and ketals are similar structures and are derived from ketones.

16. **B** Adding a second alcohol results in an acetal because protonation of the hydroxyl group of the hemiacetal is followed by the formation of a resonance-stabilized carbocation (shown with oxygen assuming the position charge). Nucleophilic attack of this highly electrophilic species followed by proton transfer gives the acetal.

17. **B** The same nucleophilic addition mechanism holds true for either aldehydes or ketones.

18. **A** Adding two molecules of alcohol results in a ketal. By analogy, a hemiketal results from the addition of one alcohol molecule.

19. **C** This reaction is similar to the one shown for acetal formation. However, in this situation, instead of a second nucleophilic addition, the reaction forms an imine from the elimination of water.

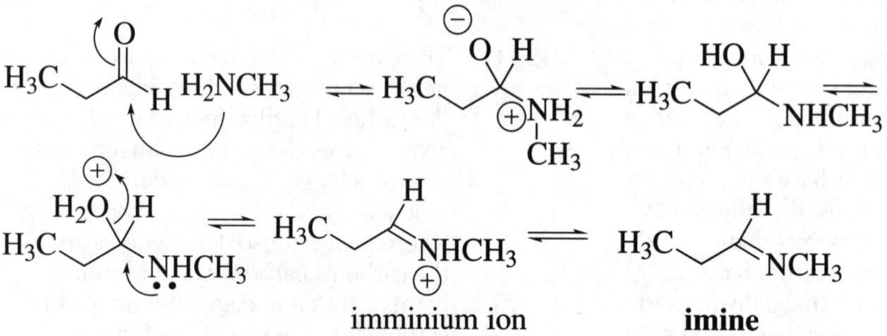

H_3C—C(=O)—H H$_2$NCH$_3$ $\rightleftharpoons$ H_3C—C(O$^-$)(H)—$^+$NH$_2$—CH$_3$ $\rightleftharpoons$ H_3C—C(HO)(H)—NHCH$_3$

H_3C—C($^+$H$_2$O)(H)—NHCH$_3$ $\rightleftharpoons$ H_3C—C(H)($^+$)—NHCH$_3$ imminium ion $\rightleftharpoons$ H_3C—C(H)=NCH$_3$ **imine**

20. **C** The second step is an elimination of water or a dehydration step. This step occurs as a lone pair from nitrogen forms a π-bond with the carbon and kicks out water.

21. **B** The reaction shown for the solution to question 19 should suggest the intermediates. The only one that makes sense is choice B, which is the intermediate after transfer of the proton. The other choices either have more than the allowable four bonds to a carbon or

do not correctly show the nucleophilic addition.

22. **B** The proton that is easiest to abstract with a base is the one that is the most acidic. In other words, the one that is easiest to remove is the one with the most stabilized conjugate base. Abstraction of the α-C proton leads to a carbanion that can be stabilized via resonance.

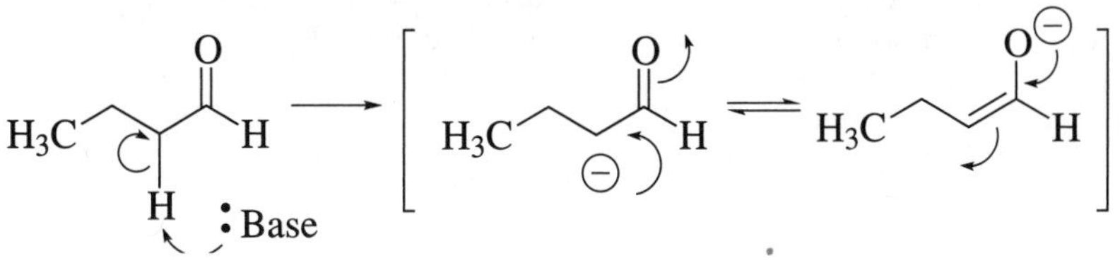

23. **C** Tautomerism is a special kind of isomerism in which two or more different molecules with identical molecular formulas are in equilibrium. Often, tautomers differ in the positions of a few bonds and a proton. Ketones, in the presence of a catalytic amount of acid or base, can tautomerize to the enol form.

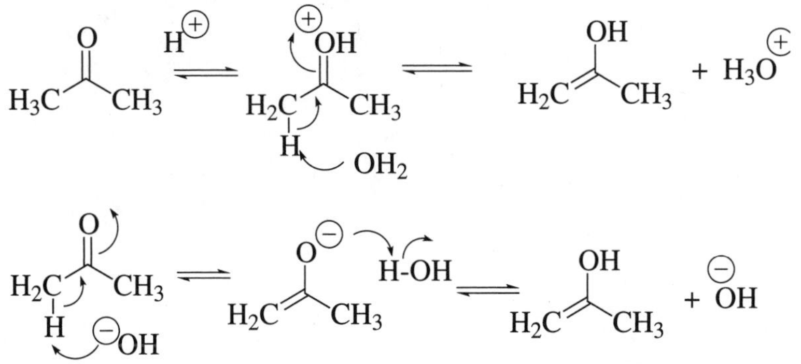

24. **D** Explained in solution to question 22.

25. **A** As a general class, the ketone form is more stable than the enol form. Remember that keto–enol tautomers are in a state of equilibrium. However, with outside influences, the keto-form makes up as much as 99% of the molecules. A notable exception is phenol; here, the molecule assumes the enol form to take advantage of the enhanced stability associated with aromaticity.

26. **B** The base removes the α-proton, and the resulting nucleophile attacks the carbon of the carbonyl of benzaldehyde, resulting in a β-hydroxy aldehyde.

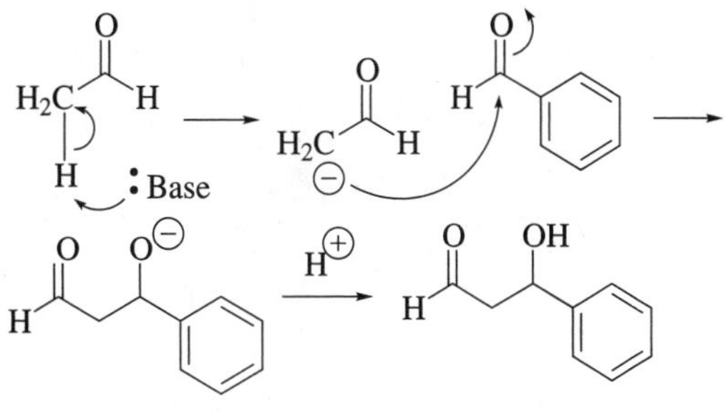

27. **B**   Explained in the preceding mechanism and in the solution to question 26.

28. **D**   The carbanion acts as a nucleophile and seeks an electrophilic center (carbon of the carbonyl). The reaction proceeds via an addition mechanism.

29. **A**   See the preceding mechanism.

30. **A**   Adding warm base leads to elimination of water (dehydration reaction of alcohols). Generally, dehydrations do not occur readily; however, in these situations, the reaction is pushed along by the fact that elimination leads to a stabilized conjugated species.

31. **A**   If base is added only to a solution of acetaldehyde, the carbanion can form as usual and then react with another molecule of acetaldehyde. Benzaldehyde does not have any α-protons; therefore, it does not form a carbanion or self-condense.

32. **C**   Ketones are considered difficult to oxidize because carbon–carbon bonds must be broken for the oxidizing to occur in this high-energy process.

Section II: The Chemistry of Oxygen-Containing Organic Compounds

Phenols and Ethers

1. A correct name for the following compound is:

$$C_6H_5\text{-}O\text{-}C(CH_3)_3$$

A. 3-phenoxytrimethane.
B. dimethyl ethyl phenyl ether.
C. *tert*-butyl phenyl ether.
D. 2,2-dimethyl ethyl phenyl ether.

2. A correct name for the following compound is:

$$H_2C=CH\text{-}O\text{-}CH=CH_2$$

A. 1,3-butene ether.
B. 1,3-butene-2-ether.
C. 1,3-diethene ether.
D. divinyl ether.

3. A correct name for the following compound is:

A. 3-bromo-1-hydroxybenzene.
B. 1-bromo-3-hydroxybenzene.
C. 3-hydroxyphenol.
D. 3-bromophenol.

4. A correct name for the following compound is:

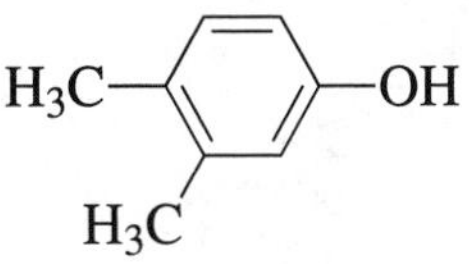

A. 3,4-dimethyl-1-hydroxybenzene.
B. 3,4-dimethylphenol.
C. 4,5-dimethyl-1-hydroxybenzene.
D. 4,5-dimethylphenol.

5. Rank the following compounds in order of increasing boiling point:

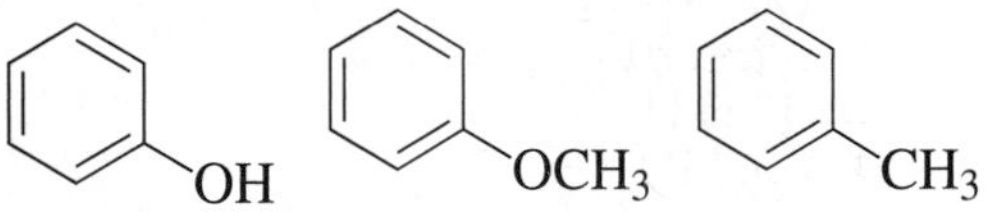

A. 1, 2, 3   **C.** 1, 3, 2
B. 3, 1, 2   **D.** 3, 2, 1

6. What is the maximum number of H-bonds that two molecules of diethyl ether can form with one another?

A. 1   **C.** 3
B. 2   **D.** None of the above

7. The BEST explanation for the difference in boiling points of ethers and phenols of the same molecular weight is:

A. more intramolecular H-bonds for phenols.
B. less intermolecular H-bonds for ethers.
C. resonance structures helping to stabilize phenols.
D. more stable sp^2 hybrid orbitals in the aromatic ring of phenols.

8. How do ethers and alcohols of similar molecular weight tend to compare in terms of water solubility?

A. Ethers are more soluble.
B. Alcohols are more soluble.
C. Both have similar solubilities.
D. Both are insoluble in water.

9. Which group forms the strongest H-bonds to water molecules?

A. Alcohols
B. Ethers
C. Phenols
D. All equally strong

10. Which one is most acidic?

A. Hexanol
B. Phenol
C. Water
D. Diisopropyl ether

11. Which statements BEST explain the difference in acidity between an alcohol and phenol of the same molecular weight?

    I.   The aromatic ring is electron-withdrawing.
    II. The phenol has low ring strain.
    III. Phenols have high H-bonding capacity.
    IV. The aromatic ring is resonance-stabilized.

    A. I and III
    B. II and III
    C. III and IV
    D. I and IV

12. Rank the following molecules in decreasing order of boiling points:

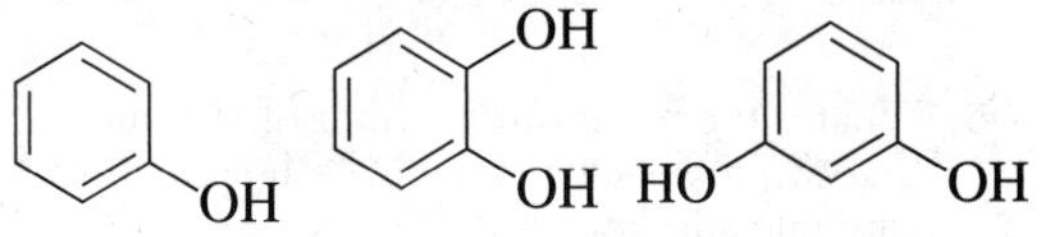

    A. 3, 2, 1
    B. 2, 3, 1
    C. 3, 1, 2
    D. 2, 1, 3

13. The bond angle between the ethyl groups in diethyl ether is closest to:

    A. 105°.
    B. 110°.
    C. 115°.
    D. 120°.

14. The highly reactive three member ring cyclic ethers are known as:

    A. epoxyranes.
    B. hydroxyranes.
    C. terpenes.
    D. oxiranes.

15. Which molecule is the most acidic?

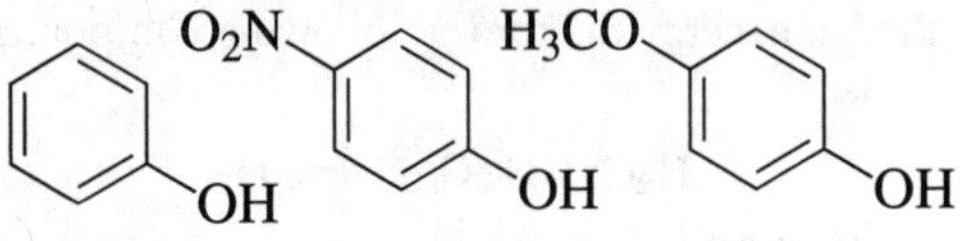

    A. I
    B. II
    C. III
    D. II and III have similar acidities.

16. What type of mechanism occurs after protonation of the ether oxygen?

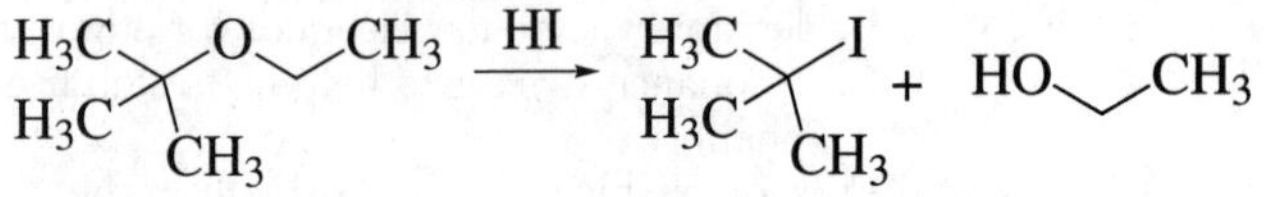

    A. S_N1
    B. S_N2
    C. E1
    D. E2

17. What type of mechanism occurs after protonation of the ether oxygen?

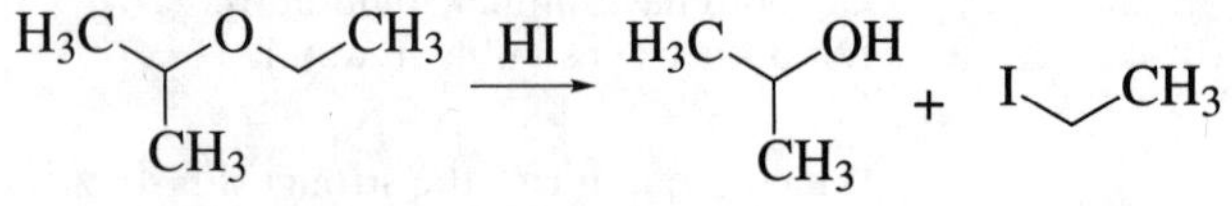

    A. S_N1
    B. S_N2
    C. E1
    D. E2

SOLUTIONS

Phenols and Ethers

1. **C** Ethers are usually called by their common names. Identify and list in alphabetic order the alkyl or other groups attached to oxygen. In this compound, a phenyl group and a *tert*-butyl group are attached to the oxygen. The *tert* (as well as the *iso* and *sec*) are not considered when determining alphabetic precedence.

2. **D** The groups attached to the oxygen are both vinyl groups. The name of this compound is divinyl ether.

3. **D** Compounds with hydroxyl groups attached to benzene are named phenols. Do **not** call these compounds hydroxybenzene. The carbon to which the hydroxyl group is attached is called carbon 1. Number the other carbons in the ring successively, giving the substituents the lowest possible numbers. Following these principles, number the ring so that the bromine is attached to carbon 3. The name of this compound is 3-bromophenol and also *m*-bromophenol.

4. **B** Number the carbons on the ring, starting with the carbon attached to the hydroxy group. Give the substituents the smallest possible numbers. Therefore, the methyl groups are numbers 3 and 4.

5. **D** Choice 3 cannot form H-bonds and is nonpolar. It has the lowest boiling point. Choice 2 has a polar nature because its electronegative oxygen can participate in polar interactions with other molecules. Choice 1, phenol, has a hydroxy group that can participate in H-bonding. Phenol, therefore, has the highest boiling point.

6. **D** Diethyl ether has a single oxygen atom that is bound to two ethyl groups. No hydrogen atoms attached to this oxygen can participate in hydrogen bonds. Although diethyl ether can H-bond to water molecules, it cannot H-bond to molecules of itself.

7. **B** Phenols have one hydroxy group and cannot form intramolecular H-bonds. The ethers cannot H-bond one another and, therefore, have fewer (actually they have none) intermolecular H-bonds to one another. The resonance structures of phenols make phenol more acidic than other alcohols. This resonance effect does not affect boiling point.

8. **C** Ethers can form H-bonds with water and, therefore, have solubilities in water that are both similar to those of alcohols of the same molecular weight and very different from those of hydrocarbons.

9. **C** Phenols form even stronger H-bonds to water because the hydrogen can be donated to water more effectively (protons are more acidic); see answer to question 10.

10. **B** Phenols have greater acidity than alcohols because of the greater delocalization of the negative charge of the phenoxide. Phenol is 10^8 times more acidic than the nonaromatic cyclohexanol.

11. **D** The aromatic ring is electron-withdrawing and allows the hydrogen atom of the hydroxy group to leave as a proton. In alcohols, alkyl groups are electron-donating and make the hydrogen less likely to leave as a proton because they destabilize the conjugate base. The enhanced stabilization of the conjugate base of phenol, owing to resonance, also explains why phenol has greater acidity than an alcohol of similar weight.

12. **A** Hydroxyl groups can more easily form intramolecular H-bonds in the *ortho* position than in the *meta* or *para* positions. These intramolecular H-bonds decrease boiling point. Therefore, compound 3 has a higher boiling point than compound 2. Compound 2, with two hydroxyl groups, has a still higher boiling point than phenol.

13. **B** The oxygen in ethers has two pairs of unbonded electrons and two groups attached to them. Water has a similar relationship, although the oxygen atom in water has only two hydrogens as substituents. The bond angle between the hydrogens in water is 105°. However, in dimethyl ether, the two methyl groups repel each other and the bond angle becomes 110°.

14. **D** When ethers are closed to form three-member ring cyclic structures, they become highly reactive **oxiranes.** These three-member cyclic ethers have a large amount of ring strain because the normal C-O-C bond angle of 109°–110° is distorted to approximately 60°. The relief of strain obtained by breaking the C-O bond provides the driving force for reactions of the oxiranes.

15. **B** Acidity is enhanced when the conjugate base is further stabilized. By placing an electron-withdrawing group on the ring, the phenoxide is further stabilized when the charge spreads over more atoms (greater delocalization). The methoxy group is electron-inducing and destabilizes the phenoxide.

16. **A** After protonation, ethanol acts as a leaving group that is leaving a relatively stable 3° carbocation, which then reacts with I$^-$ to become the *tert*-butyl iodide. The key is to recognize that the iodide ends up on the more hindered carbon. Therefore, the reaction must go through a carbocation intermediate.

17. **B** In this situation, the iodide ends up in the less hindered position, and the more hindered fragment (2-propanol) is the leaving group after protonation. This result indicates that the nucleophile has a preference for the less hindered position. Therefore, the assumption is that the reaction went through an S$_N$2.

Section II: The Chemistry of Oxygen-Containing Organic Compounds

Carboxylic Acids

1. The correct IUPAC name for the following compound is:

   A. 3-methylheptanoic acid.
   B. 4-methylheptanoic acid.
   C. 3-methyloctanoic acid.
   D. 4-methyloctanoic acid.

2. The correct IUPAC name for the following compound is:

   A. 1-chloro-2-pentanoic acid.
   B. 5-chloro-3-pentenoic acid.
   C. 1-chloro-pent-2-eneoic acid.
   D. 5-chloro-2-butenoic acid.

3. If R stands for an aliphatic hydrocarbon group, which one represents an acyl group?

4. Which statement about the boiling points of carboxylic acids it NOT true?

   A. Carboxylic acids have higher boiling points than aldehydes and ketones of similar molecular weight.
   B. The boiling points of carboxylic acids tend to increase with increasing molecular weight.
   C. Intramolecular H-bonding in carboxylic acids does not affect their boiling points.
   D. Intermolecular H-bonding in carboxylic acids has a much greater effect than intramolecular H-bonding in terms of increasing boiling point.

5. Which carboxylic acid(s) show(s) appreciable solubility in water?

   A. Methanoic acid only
   B. Methanoic to butanoic acid
   C. Methanoic to heptanoic acid
   D. Methanoic to decanoic acid

6. A correct name for the following compound is:

   A. 1-phenylmethanoic acid.
   B. 1-benzylmethanoic acid.
   C. phenyloic acid.
   D. benzoic acid.

7. $CH_3(CH_2)_4COO^-Na^+$ is the sodium salt of hexanoic acid. The water solubility of this salt is:

   A. higher than that of hexanoic acid.
   B. lower than that of hexanoic acid.
   C. completely insoluble.
   D. not predictable.

8. How many principal resonance forms does ethanoic acid have?

   A. 4   C. 2
   B. 3   D. 1

9. Any factor that stabilizes the conjugate base of a carboxylic acid more than it stabilizes the acid itself:

   A. increases the strength of the conjugate base.
   B. increases the strength of the acid.
   C. does both of the above.
   D. does neither of the above.

10. The greater acidity of carboxylic acids compared to alcohols arises primarily from:

- **A.** the electron-donating effect of the hydroxyl group.
- **B.** the electron-withdrawing effect of the carbonyl oxygen.
- **C.** the acidity of α-hydrogens of carboxylic acids.
- **D.** the resonance stability associated with the carboxylate ion.

11. Rank in order of increasing acidity:

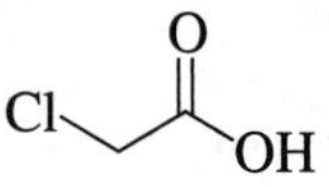

1 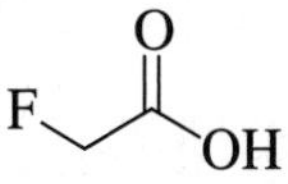   **2** 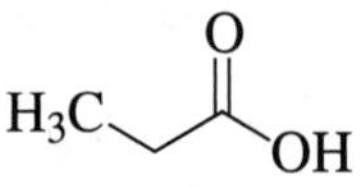   **3** 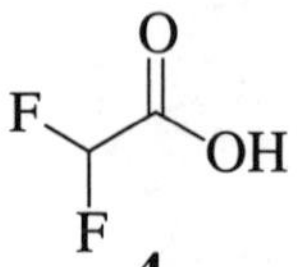   **4**

- **A.** 3, 2, 4, 1
- **B.** 1, 3, 2, 4
- **C.** 3, 2, 1, 4
- **D.** 3, 1, 2, 4

12. Of the two pairs of acids, which one is the stronger in each pair?

Pair 1

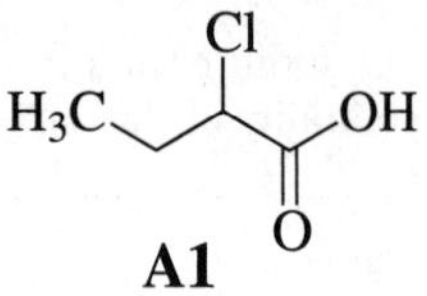

A1 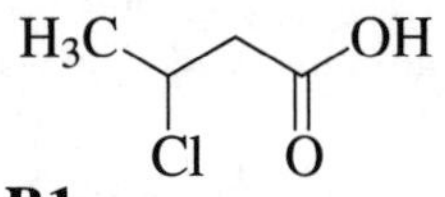   **B1**

Pair 2

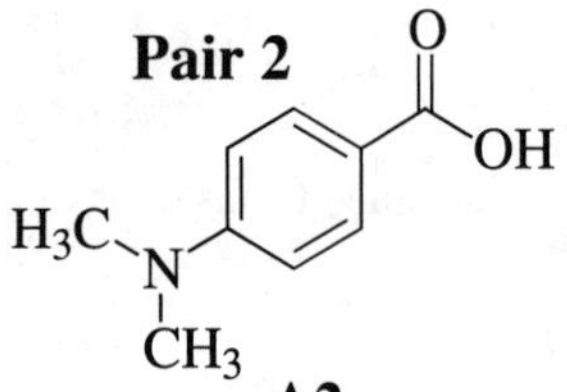

A2 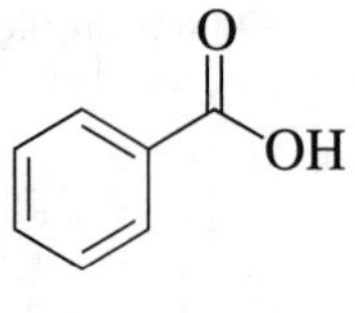   **B2**

- **A.** A1, A2
- **B.** A1, B2
- **C.** B1, B2
- **D.** B1, A2

13. When dicarboxylic acids and carboxylic acids have similar molecular weights, how do their melting points compare?

- **A.** Carboxylic acids have greater melting points.
- **B.** Dicarboxylic acids have greater melting points.
- **C.** Both acids have similar melting points.
- **D.** No consistent trend exists.

14. Rank in order of decreasing acid strength:

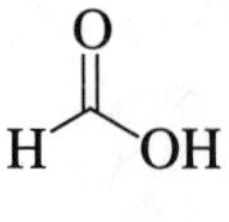

1 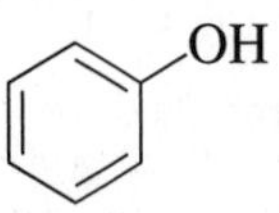   **2** 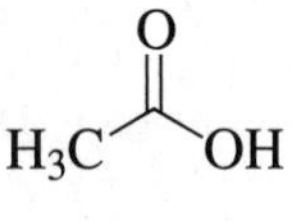   **3** 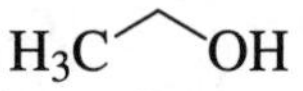   **4** 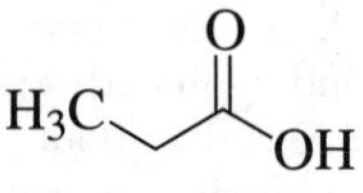   **5**

- **A.** 1, 3, 2, 5, 4
- **B.** 5, 3, 1, 2, 4
- **C.** 1, 3, 5, 2, 4
- **D.** 5, 3, 1, 4, 2

158

15. The characteristic reaction of carboxylic acids is:

A. electrophilic addition.
B. electrophilic substitution.
C. nucleophilic addition.
D. nucleophilic substitution.

16. The reactions of aldehydes or ketones and the reactions of carboxylic acids are similar in that:

A. substitution reactions tend to occur.
B. addition reactions tend to occur.
C. nucleophilic attack of the carbonyl carbon occurs.
D. two of the above occur.

Questions 17–20 refer to the following reaction:

The carboxylic acid derivative, acetyl chloride, reacts with water to form compound **A** through two main intermediates.

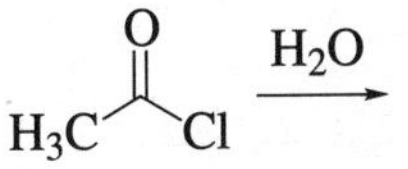

Intermediate 1 ⟶

⟶ **Intermediate 2** ⟶ **Compound A** + HCl

17. In the first step, water acts as a(n):

A. electrophile.
B. nucleophile.
C. leaving group.
D. substitution group.

18. Which structure is most likely for intermediate 1?

A. 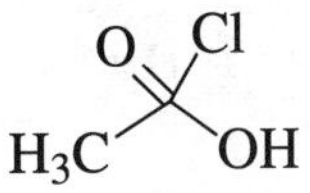

C. 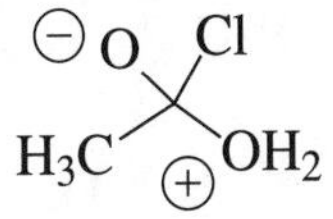

B. 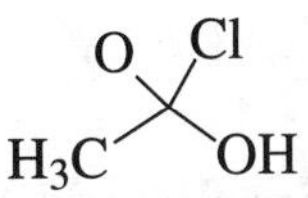

D. 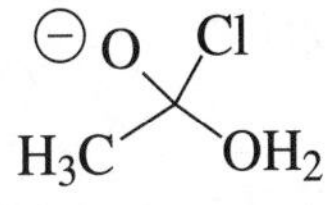

19. In the second step of the reaction, a chloride ion is lost because:

A. Cl^- is a very weak acid.
B. Cl^- is a very strong acid.
C. Cl^- is a very weak base.
D. Cl^- is a very strong base.

20. Compound A is called:

A. ethanol.
B. ethyl chloride.
C. ethyl amide.
D. ethanoic acid.

---

21. Does the following reaction occur, and why?

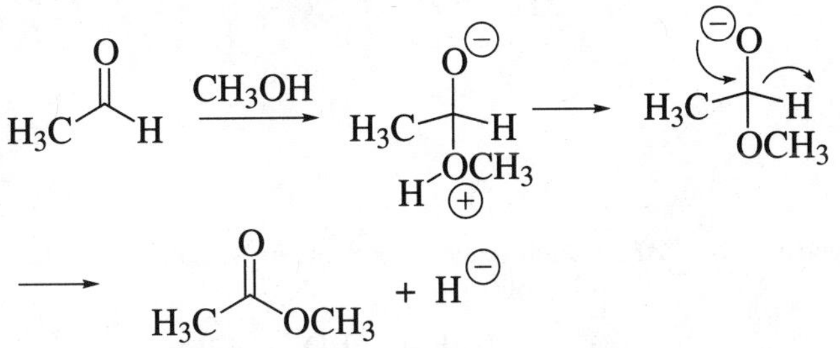

A. It occurs, because a substitution product is favored.
B. It occurs, because the hydride ion is a good leaving group.
C. It does not occur, because methanol is a very poor nucleophile.
D. It does not occur, because the hydride ion is a very powerful base.

22. Which of the following do NOT react with carboxylic acids to produce products by nucleophilic substitution mechanisms?

A. NH_2
B. CH_3OH
C. $SOCl_2$ or PCl_5
D. All of the above act as nucleophiles in substitution reactions.

23. When propanoic acid reacts with methanol that is labeled with ^{18}O in acidic solution, the ester formed is:

A. 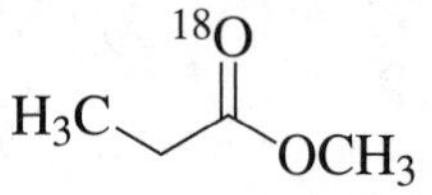
$$H_3C \underset{^{18}O}{\overset{}{\diagup}} OCH_3 \;+\; H_2O$$

B. 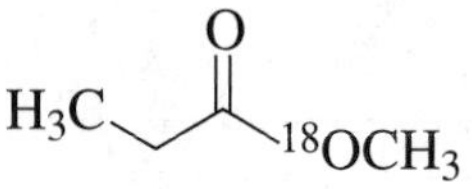
$$H_3C \overset{O}{\diagup} {}^{18}OCH_3 \;+\; H_2O$$

C. 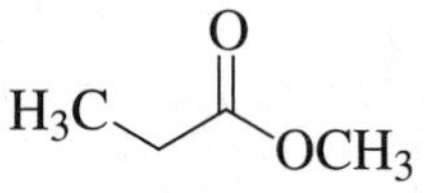
$$H_3C \overset{O}{\diagup} OCH_3 \;+\; H_2{}^{18}O$$

D. **All of the above are formed** in equal amounts

24. What is the product of the following esterification reaction?

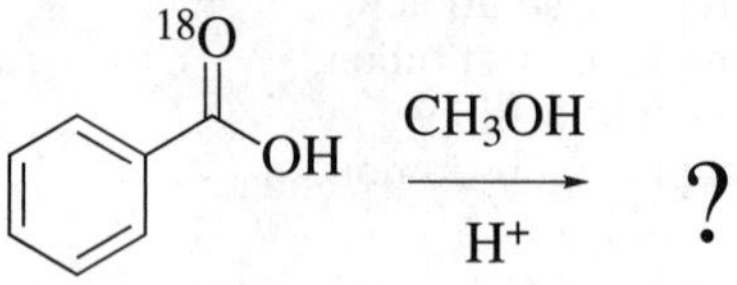

$$\text{(benzoic acid, }{}^{18}O\text{ carbonyl)} \xrightarrow[H^+]{CH_3OH} \;\; ?$$

A. 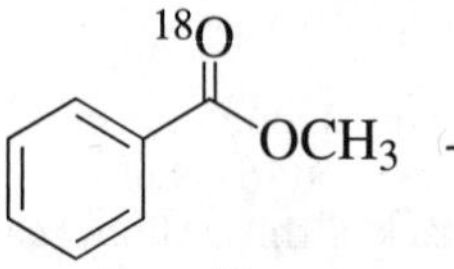
$$\text{(benzoate) } OCH_3 \;+\; H_2O \quad (\text{carbonyl } {}^{18}O)$$

B. 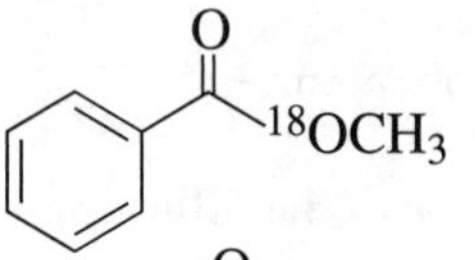
$$\text{(benzoate) } {}^{18}OCH_3 \;+\; H_2O$$

C. 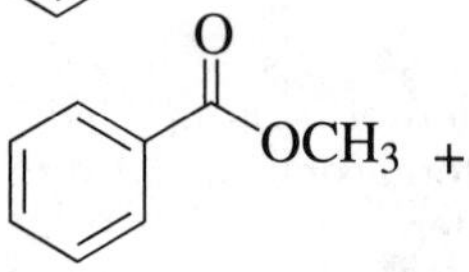
$$\text{(benzoate) } OCH_3 \;+\; H_2{}^{18}O$$

D. **A and C**

25. Which acid(s) decarboxylate(s) when heated to 100°–150°C?

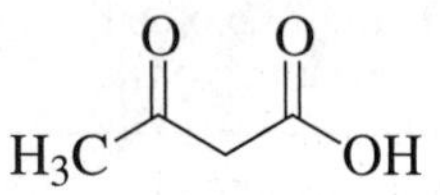

$$H_3C \overset{O}{\diagup}\overset{O}{\diagup} OH \qquad H_3C \overset{O}{\diagup}\diagdown\overset{O}{\diagup} OH \qquad HO \overset{O}{\diagup}\overset{O}{\diagup} OH$$

I 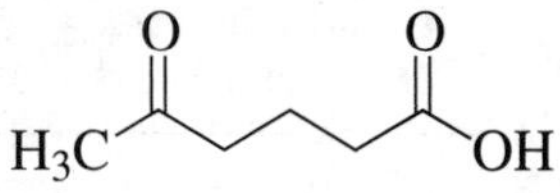   **II** 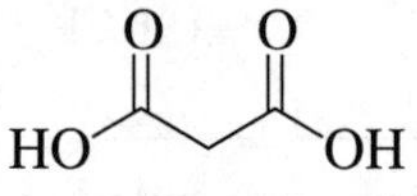   **III**

A. I
B. I and II
C. II and III
D. I and III

26. Compounds that decarboxylate readily are known as:

   I.   gamma-diacids.
   II. beta-diacids.
   III. dicarboxylic acids.
   IV. beta-ketoacids.
   V. gamma-aldacids.

A. I and V
B. III and IV
C. III and V
D. II and IV

27. Which reagent(s) reduce(s) ethanoic acid to ethanol?

A. $LiAlH_4/Et_2O$, H_2O
B. $NaBH_4$
C. Both
D. Neither

SOLUTIONS

Carboxylic Acids

1. **D** Identify the longest carbon chain attached to the carboxylic acid. Number, starting from the carbonyl carbon of the acid. Therefore, the methyl group is at position 4. Drop the "e" from the base name and replace with "oic acid."

2. **B** Number the compounds, starting from the carbonyl carbon of the acid. The acid takes priority over the double bond. The double bond starts at carbon 3 and the Cl is at carbon 5. The base name is pentene. Because the carboxylic acid is always the terminal group, it is not necessary to split up the name as in choice C.

3. **D** Choice A is a carbonyl, B is a carboxylic acid, and C is an ether.

4. **C** Intramolecular H-bonding tends to decrease boiling points. Instead of the molecule forming intermolecular H-bonds, H-bonding occurs within the same molecule. Therefore, there are no increased intermolecular forces to affect boiling point.

5. **B** The first four carboxylic acids, containing one to four carbons, show appreciable solubility in water. As the number of carbons increase, the hydrophobic nature of the carbon chain takes charge and the water solubility starts to decrease.

6. **D** This acid is similar to benzaldehyde and benzyl alcohol from previous sections.

7. **A** Sodium and potassium salts of the majority of carboxylic acids, even of long-chain carboxylic acids, are soluble in water. Salts of long-chain carboxylic acids are a major component of soap. Without the counterion, the long-chain carboxylic acids would be fully insoluble in water.

8. **B** The diagram shows the three main resonance structures for ethanoic acid.

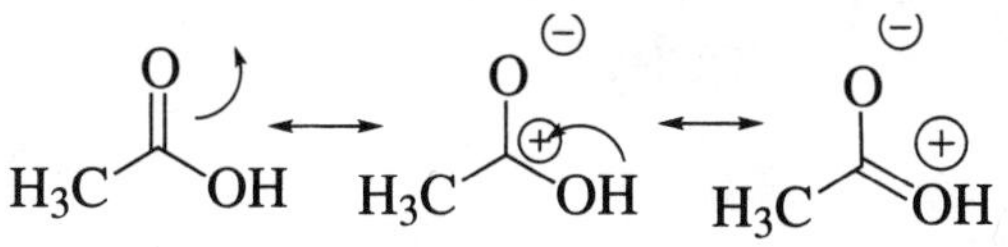

9. **B** Remember that acids are proton donors; any factor that stabilizes the conjugate base shifts the equilibrium toward proton donation. If the conjugate base is highly reactive (basic), it grabs a proton and reverts to the acid it was at first.

10. **D** The conjugate base of a carboxylic acid is stabilized by resonance. The conjugate base of an alcohol (alkoxide) does not benefit from this type of stabilization and is consequently more reactive (basic); therefore, the alcohol is a weaker acid.

11. **D** Three major factors influence the acidity of substituted carboxylic acids. Electronegative groups attached to the molecule stabilize the conjugate base by helping to delocalize the charge. The closer these groups are to the carbonyl and the more groups there are, the greater the acidity of the acid. F is more electron-withdrawing than Cl, which is more electron-withdrawing than H.

12. **B** In Pair 1 choice A1 is a stronger acid because the Cl group is closer to the carboxyl group. For Pair 2, B2 is the stronger acid because the amino is electron-donating (actually it is a base itself) and it destabilizes the conjugate base.

13. **B** The dicarboxylic acids (aliphatic) tend to have much higher melting points than the monocarboxylic acids of similar molecular weight. For example, the molecule HOOC-COOH has a melting point of 189°C whereas the molecule CH_3CH_2COOH has a melting point of −6°C.

14. **C** Remember that alkyl groups are electron-donating and they destabilize the conjugate base. Therefore, as the carbon chain increases, the acidity decreases. All of the carboxylic acids are stronger acids than phenol, which is a stronger acid than ethanol.

15. **D** The characteristic reaction of carboxylic acids is nucleophilic substitution. Although the initial step in both nucleophilic addition and substitution is

the same (nucleophilic attack of the carbonyl carbon to form a tetrahedral intermediate), the later steps differ. In substitution reactions of acids, after formation of the tetrahedral intermediate, a leaving group is ejected, allowing for regeneration of the carbonyl. Therefore, the extent to which these substitution reactions can occur depends on how good a leaving group the ejected molecule is. Aldehydes and ketones do not undergo substitution because a hydride (H^-) and an alkyl anion are not good leaving groups. Both are very strong bases.

16. **C**   Discussed in the solution for question 15

---

17. **B**   This situation is an example of nucleophilic substitution. In the first step, the nucleophile (water) attacks the electrophile to form a tetrahedral intermediate, followed by pushing out the chloride to give the acid.

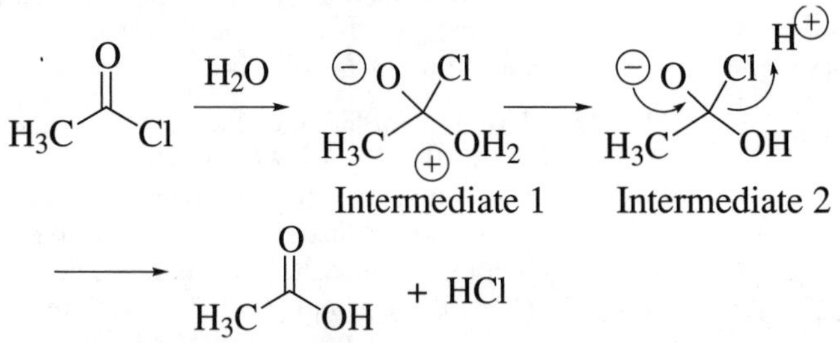

18. **C**   The intermediate is shown in the preceding mechanism. The incorrect choices have more than four bonds to a carbon, predict that an incorrect molecule has attacked the carbonyl, or give incorrect charges.

19. **C**   Weak bases (conjugate bases of strong acids) are good leaving groups because they can stabilize the negative charge they assume after they depart and are therefore less reactive. Cl^- is the conjugate base of the strong acid HCl and is therefore a weak base.

20. **D**   Compound A is the substitution product, ethanoic acid.

21. **D**   In the diagrammed reaction, an aldehyde undergoes nucleophilic substitution. This substitution is unlikely because aldehydes do not contain good leaving groups. The hydride (H^-) is a very strong base and therefore a poor leaving group. An addition product is more likely.

22. **D**   All of the choices can act as nucleophiles and attack the carbonyl carbon of the carboxylic acid.

---

23. **B**   The following mechanism shows the movement of the labeled alcohol. It is an acid-catalyzed esterification reaction.

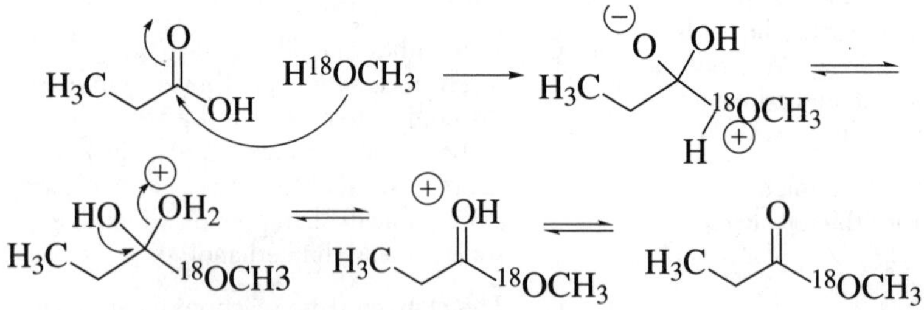

24. **D**   The mechanism shown in the solution for question 23 indicates that the carbonyl oxygen is not displaced in the esterification. However, it is also

impossible to label one oxygen and have it keep its label as a carbonyl. The resonance between the two oxygens results in an exchange of the hydrogen from one oxygen to the next. Therefore, some of the carbonyl and some of the water will be labeled after esterification.

25. **D** Beta-diacids and beta-ketoacids decarboxylate readily when heated. This decarboxylation occurs because a carboxylic acid with a carbonyl in the beta position destabilizes the acid. When the acid is protonated, a six-member H-bonded structure can form and initiate the decarboxylation. CO_2 is given off as a gas, and therefore, this event is entropically favorable. This event occurs only when the carbonyl groups are in the beta position because this intermediate is not readily formed otherwise.

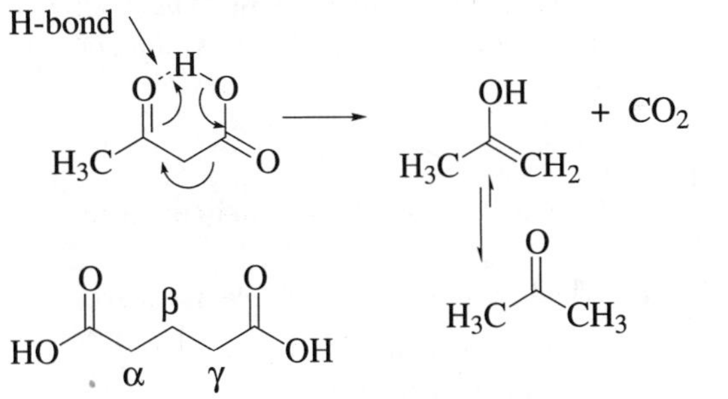

26. **D** Discussed in the solution for question 25

27. **A** Lithium aluminum hydride is a strong reducing agent for all carbonyl compounds. Sodium borohydride is a weaker reducing agent and cannot reduce a carboxylic acid.

Section III: Organic Molecules of Biological Importance

Amines

1. The correct IUPAC name for the following compound is:

    H_3C ‚‚‚ NH_2
    CH_3

    A. 2-methyl-1-butanamine.
    B. 3-methyl-4-butanamine.
    C. isobutylamine.
    D. N-methylbutanamine.

2. The correct IUPAC name for the following compound is:

    H_3C ‚‚ N ‚‚ CH_3
        H

    A. 2-propanamine.
    B. 1-propanamine.
    C. N-ethylmethanamine.
    D. N-methylethanamine.

3. The correct IUPAC name for the following compound is:

    H_3C ‚‚‚ N ‚‚ CH_3
            CH_3

    A. N-propyl-N-ethyl-1-methanamine.
    B. N-methyl-N-propyl-1-propanamine.
    C. N-ethyl-N-methyl-1-propanamine.
    D. N,N,N-ethylmethylpropylamine.

4. The correct common names for the following compounds are (from left to right):

    H_2N ‚‚‚ (ring)   (ring N)   (ring N, N)   (ring N, N, N, N, H)

    1   **2**   **3**   **4**

    A. pyridine, aniline, purine, pyrimidine.
    B. aniline, pyrimidine, pyridine, purine.
    C. aniline, pyridine, purine, pyrimidine.
    D. pyridine, aniline, pyrimidine, purine.

5. Which compound has the highest boiling point?

    $CH_3CH_2CH_3$ CH_3CH_2OH $CH_3CH_2NH_2$
        I   **II**   **III**

    A. I
    B. II
    C. III
    D. II and III have equivalent boiling points.

6. Consider a 1°, 2° and 3° amine, all of equivalent molecular weight. Which amine is most likely to have the lowest boiling point?

    A. 1° amine
    B. 2° amine
    C. 3° amine
    D. Not enough information to determine

7. Which amine is NOT soluble in water?

    A. Methylamine
    B. Dimethylamine
    C. Trimethylamine
    D. All are water-soluble.

8. A sample of pure amine molecules is found to possess no intermolecular H-bonding. This sample is most likely:

    A. 1° amines.
    B. 2° amines.
    C. 3° amines.
    D. all of the above.

9. In the pure gas phase, which basicities of the amines are correctly represented (high to low)?

    A. $NH_3 > NH_2CH_3 > NH(CH_3)_2 > N(CH_3)_3$
    B. $N(CH_3)_3 > NH(CH_3)_2 > NH_2CH_3 > NH_3$
    C. $NH_3 > NH_2CH_3 > NH(CH_3)_2 = N(CH_3)_3$
    D. $N(CH_3)_3 = NH(CH_3)_2 > NH_2CH_3 > NH_3$

10. Rank the following compounds in order of increasing base strength:

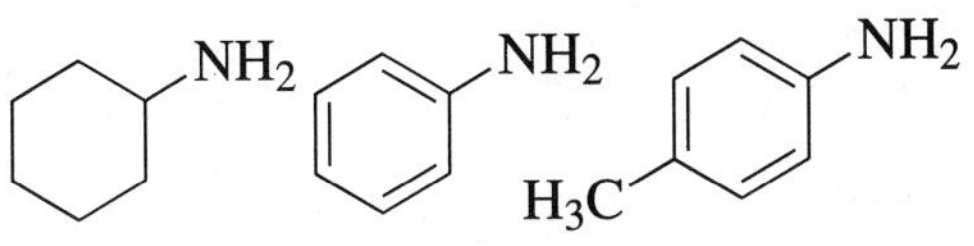

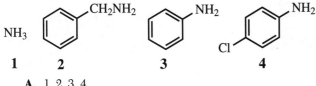

1   **2**   **3**

A. 1, 2, 3
B. 3, 2, 1
C. 2, 3, 1
D. 2, 1, 3

11. Rank the following compounds in order of increasing base strength:

NH$_3$ CH$_2$NH$_2$ NH$_2$ NH$_2$

Cl

1   **2**   **3**   **4**

A. 1, 2, 3, 4
B. 4, 2, 3, 1
C. 1, 4, 3, 2
D. 4, 3, 1, 2

12. Which one correctly depicts a quaternary ammonium ion?

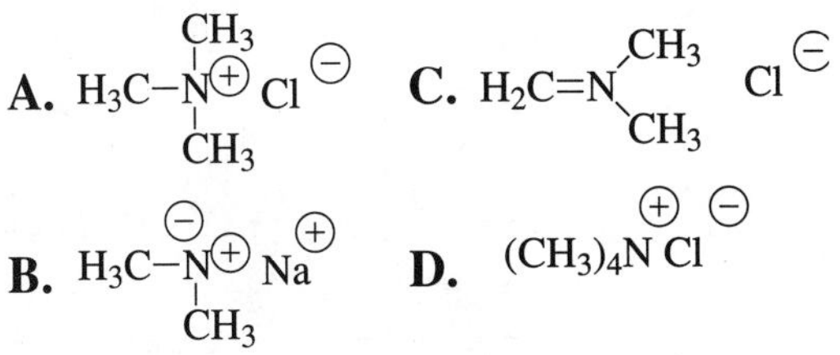

A. $H_3C-\overset{\underset{|}{CH_3}}{\overset{+}{N}}-CH_3 \quad Cl^{-}$

C. $H_2C=N\overset{CH_3}{\underset{CH_3}{}} \quad Cl^{-}$

B. $H_3C-\overset{\underset{|}{CH_3}}{\overset{+}{N}^{-}} \quad Na^{+}$

D. $(CH_3)_4N^{+} \; Cl^{-}$

13. The one that does NOT tend to react with acyl chlorides to form amides is:

A. ammonia.
B. 1° amines.
C. 2° amines.
D. 3° amines.

Questions 14–16 refer to the following reaction:

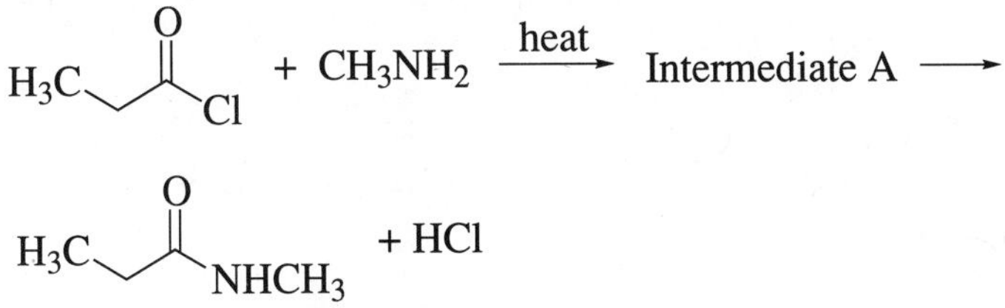

$$H_3C-\overset{O}{\underset{}{C}}-Cl + CH_3NH_2 \xrightarrow{\text{heat}} \text{Intermediate A} \longrightarrow$$

$$H_3C-\overset{O}{\underset{}{C}}-NHCH_3 + HCl$$

14. The overall reaction is considered:

A. electrophilic addition.
B. nucleophilic addition.
C. electrophilic substitution.
D. nucleophilic substitution.

...mediate A is most likely:

A. 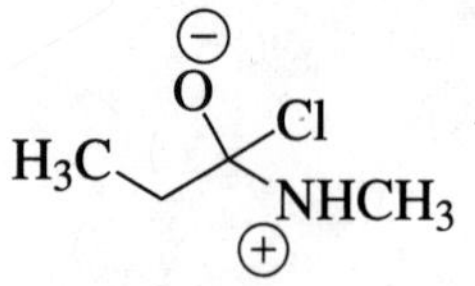 H_3C—C with O^- (top), Cl, and $NHCH_3$ ($+$) substituents

C. H_3C—C with O^- (top), Cl, and NH_2CH_3 ($+$) substituents

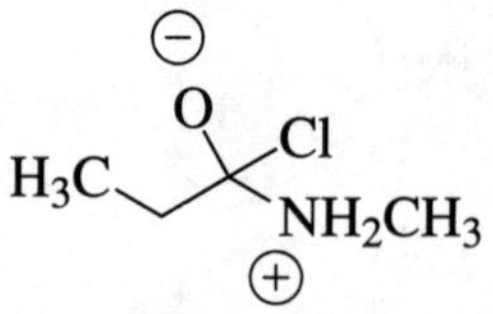

B. 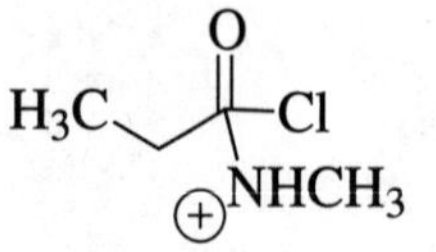 H_3C—C(=O)—Cl with $NHCH_3$ ($+$) substituent

D. 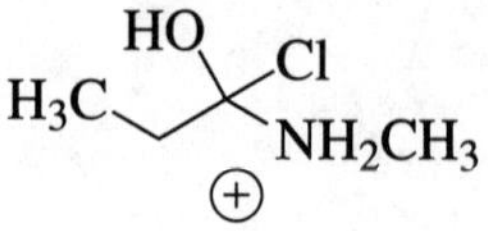 H_3C—C with HO (top), Cl, and NH_2CH_3 ($+$) substituents

16. The major product of this reaction is a(n):

  A. 3° amine.
  B. 2° amine.
  C. ester.
  D. amide.

17. When ammonia is added to an alkyl halide, in the presence of base:

  A. primary amines form.
  B. amides form.
  C. nitrated alkyl halides form.
  D. quaternary ammonium salts form.

18. Alkylation of amines proceeds by:

  A. electrophilic addition.
  B. nucleophilic addition.
  C. electrophilic substitution.
  D. nucleophilic substitution.

19. The major product of the following reaction is:

$CH_3CH_2Cl + (CH_3)_2 NH \rightarrow$?

  A. 1° amines.
  B. 2° amines.
  C. 3° amines.
  D. amides.

166

SOLUTIONS

Amines

1. **A** In the IUPAC system, primary amines are named by dropping the "e" from the base name of the carbon chain and replacing it with the suffix "amine." Number from the carbon attached to the N. In this situation, the carbon bearing the methyl group is number 2. The 1 in the name refers to the position of the amine.

2. **D** 2° and 3° amines are named by using the base name of the longest alkyl group attached to the N. The other groups are designated by their alkyl group names, with an "N" to indicate the position of attachment on N of the parent compound. In this question, an ethyl group is the largest alkyl group, with a methyl group attached to the nitrogen.

3. **C** The largest alkyl group is propane; therefore, the compound is a propanamine. The substituents attached to the nitrogen are an ethyl and a methyl group. Give the substituents in alphabetic order. Once again, *N* indicates that the substituents are attached to the nitrogen, and the 1 refers to the position of the nitrogen on the chain.

4. **D** Aniline is an aminobenzene. Pyridine is an aromatic amine. Pyrimidine and purine are also aromatic heterocyclic bases that make up the cyclic backbone of DNA and RNA. Adenine and guanine are purines whereas cytosine, thymine, and uracil are pyrimidines.

5. **B** Amines are fairly polar compounds, but they boil at temperatures lower than those of alcohols of similar chain length and structure. Amines do have higher boiling points than carbonyl compounds and alkanes because they possess both H-bond donors and acceptors.

6. **C** The 3° amines do not form hydrogen bonds to one another. They can, however, form H-bonds to water; therefore, all low–molecular weight amines are soluble in water. The 1° and 2° amines can form H-bonds to themselves and therefore have higher boiling points than 3° amines.

7. **D** All of these small, low–molecular weight amines are water-soluble and are capable of forming H-bonds with water.

8. **C** Discussed in the solution to question 6.

9. **B** Electron-releasing groups attached to nitrogen increase the basicity of amines by stabilizing the positive charge gained after the nitrogen accepts a proton (Brønsted base) or donates its lone pair (Lewis base). Therefore, the amine with the most alkyl groups is the most basic.

10. **C** The aromatic amines are weaker bases than the nonaromatic cyclic amines because the lone pair electrons on N are delocalized over the aromatic ring. This delocalization stabilizes the compound but also decreases the ability of the nitrogen to act as a base. Compound 3 is a stronger base than aniline because of the electron-releasing substituent on the ring.

11. **D** Of the aromatic amines, the weakest base is the one with the electron-withdrawing substituent: (4) is the weakest followed by aniline (3). Compound 2 is the strongest base because its electrons are not delocalized effectively into the aromatic ring, and it is therefore a primary amine, and its pK_b is slightly lower than that of ammonia.

12. **D** A compound containing nitrogen that has four carbon groups attached and is associated with a salt anion is called a quaternary ammonium salt. Examination of the structures of the choices shows that only D satisfies the conditions of the definition.

13. **D** Ammonia, 1° and 2° amines all react rapidly with acyl chlorides to form amides. The reaction proceeds through a nucleophilic substitution mechanism. Because 3° amines already have three groups bound to nitrogen, they tend not to react through this mechanism to form amides.

14. **D** The mechanism for the following reaction is:

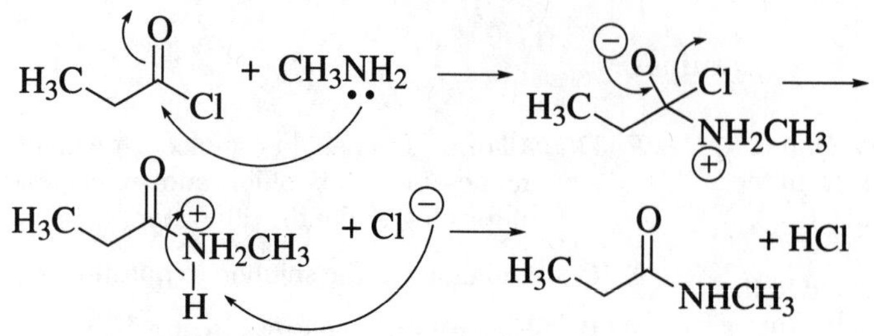

15. **C** See the mechanism given for the solution to question 14.

16. **D** The product resulting from the preceding reaction is an amide.

17. **A** Salts of 1° amines can be prepared from ammonia and alkyl halides through a nucleophilic substitution reaction. Treating the ammonium salt with base gives a 1° amine.

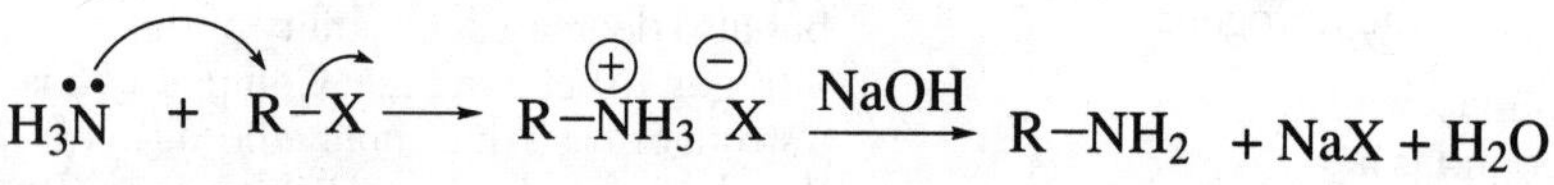

18. **D** Adding alkyl groups to amines is called alkylation of amines. It proceeds through the following nucleophilic substitution mechanism:

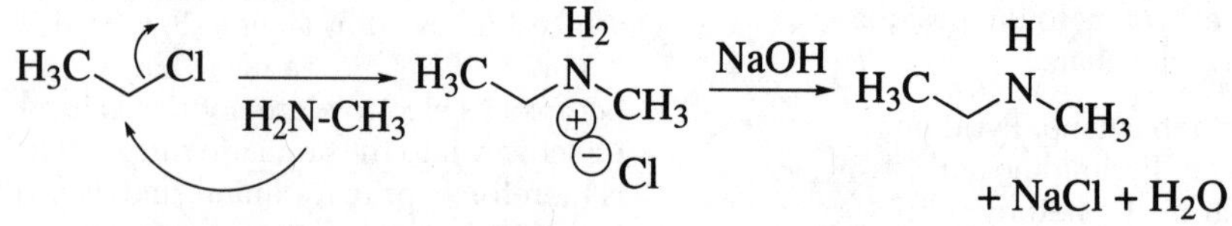

19. **C** This reaction proceeds through a nucleophilic substitution reaction to produce a 3° amine product. The mechanism is the same as the one shown in the solution to question 18.

Section III: Organic Molecules of Biological Importance

Amino Acids and Proteins

1. Amino acids can be divided into different groups depending on the structures of their side chains. Which group does NOT belong?

   A. Neutral, nonpolar
   B. Non-neutral, nonpolar
   C. Basic
   D. Acidic

2. The predominant form of an amino acid in solution depends on the:

   I.   pH of the solution
   II. nature of the amino acid.
   III. concentration of the amino acid.

   A. I
   B. II
   C. I and III
   D. I and II

3. Serine and threonine contain hydroxyl groups in their side chains. At pH 6.0 to 7.0, these amino acids are considered:

   A. acidic.
   B. basic.
   C. neutral, nonpolar.
   D. neutral, polar.

4. The amino acids leucine and isoleucine have uncharged, nonpolar side chains. At pH 6.0, these amino acids are considered:

   A. neutral.
   B. basic.
   C. acidic.
   D. undetermined.

5. Amino acids are BEST considered:

   I.   acidic.
   II. basic.
   III. zwitterions.
   IV. amphoteric.

   A. I and III
   B. II and IV
   C. II and III
   D. III and IV

6. In dipolar ionic form, the measured acidity of a neutral amino acid refers to the:

   A. acidity of the ammonium ion.
   B. acidity of the carboxylate ion.
   C. acidity of the side-chain groups.
   D. acidity of the carboxylic-acid group.

7. The concentration of hydrogen ions in a solution in which an amino acid does NOT migrate under the influence of an electric field is called the amino acid's:

   A. dipolar point.
   B. endpoint.
   C. isoionic point.
   D. isoelectric point.

Questions 8–11 refer to the amino acid alanine, which is a neutral, nonpolar amino acid. The titration of alanine is as follows:

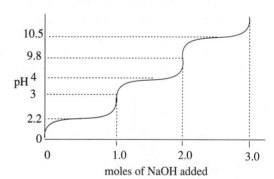

8. A solution of alanine exists in a 100% dipolar form at the pH of the:

   A. pK_a of the carboxylic acid.
   B. pK_a of the amine.
   C. pK_a of the α carbon hydrogen.
   D. pI.

9. At which pH does alanine NOT migrate in an electric field?

   A. 2.3   C. 6.0
   B. 4.0   D. 9.7

10. At which pH does alanine exist in a mixture of 50% zwitterionic and 50% fully deprotonated forms?

   A. Point A
   B. Point B
   C. Point C
   D. Point D

11. Consider a fully protonated sample of glycine. How many equivalents of sodium hydroxide must be added to reach the pK_a of the proton of the carboxylic acid?

   A. 0.5
   B. 1.0
   C. 1.5
   D. 2.0

Questions 12–15 refer to the following titration curve for an unknown amino acid:

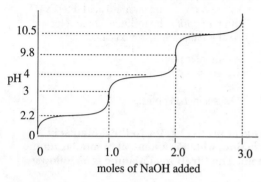

moles of NaOH added

12. The side chain of this amino acid is most likely:

   A. neutral, nonpolar.
   B. acidic.
   C. basic.
   D. neutral, polar.

13. The pH value of 4 is equivalent to the pK_a for the:

   A. carboxylic acid proton.
   B. amino group proton.
   C. side-chain group proton.
   D. ammonium group proton.

14. What is the net charge on the amino acid molecules at pH 10.5?

   A. All molecules are −2.
   B. 50% are −1, 50% are −2.
   C. All molecules are −1.
   D. 50% are neutral, 50% are −1.

15. How many equivalents of base are needed to reach the pK_a of the second most acidic proton on the molecule?

   A. 0.5
   B. 1.0
   C. 1.5
   D. 2.0

Questions 16–18 refer to the basic amino acid lysine, with pK_a = 12.5 for its side chain.

16. If the amino acid is placed in an electric field at pH 7.0:

   A. it migrates toward the anode.
   B. it migrates toward the cathode.
   C. it does not migrate because the amino acid is at its isoelectric point.
   D. 50% migrate to the anode, 50% to the cathode.

17. At pH 13, the majority of lysine molecules

   A. migrate toward the anode.
   B. migrate toward the cathode.
   C. the majority of lysine molecules do not migrate because the amino acid is at its isoelectric point.
   D. 50% will migrate to the anode, 50% to the cathode.

18. In strongly acidic solution, lysine exists as a(n):

   A. anion.
   B. dianion.
   C. cation.
   D. dication.

19. The pI for lysine is closest to:

   A. pH 4.0.
   B. pH 7.2.
   C. pH 10.8.
   D. pH 12.5.

20. Peptide bonds can be classified as:

   A. amines.
   B. amides.
   C. carboxylic acids.
   D. esters.

21. What is released when peptide bonds form from the condensation of amino acids?

   A. Hydronium ions
   B. Amines
   C. Water
   D. Hydrogen gas

Questions 22–24 refer to the hydrolysis of the following dipeptide:

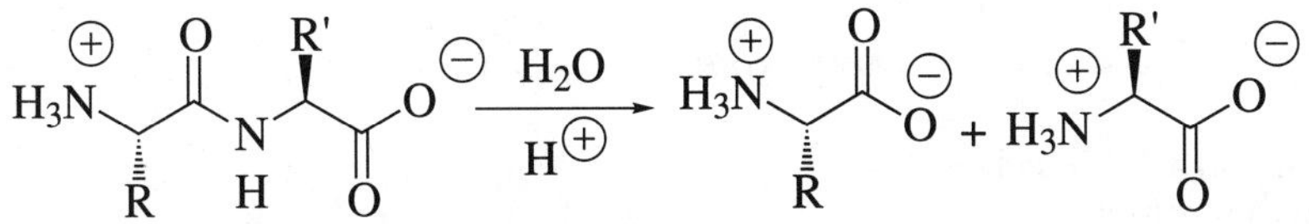

22. The acid acts to:

  A. create a 1° amine.
  B. form hydronium ions, which attack the carbonyl.
  C. cleave the peptide bond.
  D. protonate the carbonyl oxygen and activate the carbonyl for attack.

23. Water acts as a(n):

  A. catalyst.
  B. electrophile.
  C. nucleophile.
  D. leaving group.

24. The leaving group in this acidic hydrolysis is:

  A. the carbonyl end of an amino acid.
  B. the amino end of an amino acid.
  C. water.
  D. a protonated alcohol.

Questions 25–27 refer to the following molecule:

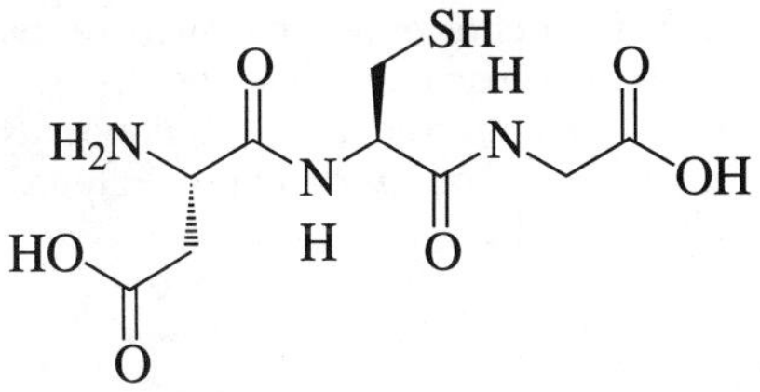

25. This molecule is classified as a:

  A. polypeptide.
  B. dipeptide.
  C. tripeptide.
  D. pentapeptide.

26. The pI of this molecule is most likely:

  A. equal to 7.
  B. greater than 7.
  C. less than 7.
  D. unable to be determined.

27. This molecule BEST demonstrates:

  A. 1° structure.
  B. 2° structure.
  C. 3° structure.
  D. quaternary structure.

28. Which does NOT relate to secondary structure in proteins?

  A. The formation of β sheets
  B. Intramolecular H-bonding
  C. The formation of α-helices
  D. Amino acid sequence

29. Which bonds do NOT stabilize the 3° structure of proteins?

  A. H-bonds
  B. Van der Waals forces
  C. Covalent and ionic bonds between side chains
  D. Peptide bonds

30. A monomeric enzyme is partially denatured by heat. Which is (are) likely to be affected by this change?

  I.   1° structure
  II. 2° structure
  III. 3° structure
  IV. Quaternary structure

  A. I
  B. II and III
  C. III and IV
  D. I, II, III, and IV

31. Collagen is comprised of three helical proteins twisted about one another and held together by hydrogen bonds. When collagen is boiled in water, gelatin forms. Gelatin consists of single-protein helices. What is the highest level of structure that has been disturbed by heating collagen?

  A. 1° structure
  B. 2° structure
  C. 3° structure
  D. Quaternary structure

SOLUTIONS

Amino Acids and Proteins

1. **B** The four groups of amino acids based on their side chains are the neutral nonpolar, neutral polar, acidic, and basic groups.

2. **D** The form of an amino acid depends on the pH and the nature of the amino acid. At very acidic conditions, amino acids are fully protonated. As the pH is increased, the amino acid is deprotonated, with the most acidic protons (lowest pK_a) leaving first. The form of the amino acid also depends on the nature of the molecule because the 20 common amino acids all have different side chains. Some of these side chains are acidic, others are basic, and still others are neutral (polar and nonpolar). Concentration of the amino acid has no effect on the form of the molecule.

3. **D** Hydroxy groups are not protonated or deprotonated at pH 7, but they are polar. At pH 7, the acid is deprotonated and the amine is protonated; therefore, the molecule has an overall charge of 0. Following is a useful fact: pK_a for carboxylic acids is approximately 2.3 and for ammonium ions is 9.7. Therefore, at pH 6 to 7, the acid is fully converted to a carboxylate and the amine is protonated (ammonium).

4. **A** This question follows the reasoning from the solution to question 3. Because the carboxyl group is deprotonated and the

amino group is fully protonated, the molecules are neutral.

5. **D** All 20 of the common amino acids contain a COOH group and a $^+NH_3$ when fully protonated at low pH. As the pH is raised, the protons are first lost from the carboxylic acid (50% lost at pH 2.3) and then from the ammonium (50% lost at pH 9.7). Protons can also be lost from acidic or basic side-chain groups. The result of these proton losses are zwitterions or dipolar ions. These terms simply mean that amino acids can both gain and lose protons and therefore are also considered amphoteric (can act as both an acid and a base). Choices I and II are correct, but they are not the best choices.

6. **A** Acids are proton donors. In dipolar form, the carboxyl group is deprotonated and does not have a proton to donate. Only the ammonium ion ($^+NH_3$) group has an available proton. Therefore, measuring the acidity of dipolar neutral amino acids is similar to measuring the acidity of the ammonium ion.

7. **D** The term defined is the isoelectric point. The other choices are imaginary terms. The isoelectric point describes the pH at which an amino acid is neutral, because neutral compounds do not migrate in electric fields.

8. **D** At pI, the amino acid must be neutral and would exist in the 100% dipolar form. The diagram describes the titration of alanine with base. (R=CH$_3$)

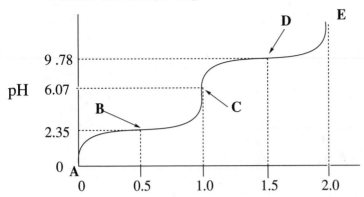

moles of NaOH added

A = 100% $\overset{\oplus}{H_3N}\overset{H}{\underset{R}{\rule{0pt}{0pt}|}}COOH$

B = 50% $H_3N\overset{H}{\underset{\overset{\oplus}{R}}{|}}COOH$ 50% $H_3N\overset{H}{\underset{\overset{\oplus}{R}}{|}}COO^{\ominus}$

C = 100% $H_3N\overset{H}{\underset{\overset{\oplus}{R}}{|}}COO^{\ominus}$

D = 50% $H_3N\overset{H}{\underset{\overset{\oplus}{R}}{|}}\overset{\ominus}{COO}$ 50% $H_2N\overset{H}{\underset{R}{|}}COO^{\ominus}$

E = 100% $H_2N\overset{H}{\underset{R}{|}}COO^{\ominus}$

9. **C** The average of the pK_a's of the monocationic form and the neutral form gives the isoelectric point or the endpoint of the titration. At this point, 100% of the amino acids are in a zwitterionic or dipolar form.

10. **D** See the preceding titration curve for the solution to question 8.

11. **A** See the preceding titration curve for the solution to question 8. The curve for alanine is similar to that of glycine.

12. **B** Note the three flat areas of the curve, which indicates three pK_a values (three acidic protons). The first occurs at a pH that is indicative of a carboxyl group. The last is indicative of an ammonium. The middle pK_a occurs well below pH 7, which indicates that the side chain is most likely an acidic side chain.

13. **C** The side chain is acidic and therefore is probably a carboxyl group (aspartic acid or glutamic acid).

14. **B**   The following diagram indicates the species at each point:

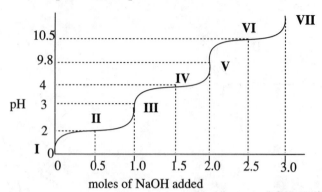

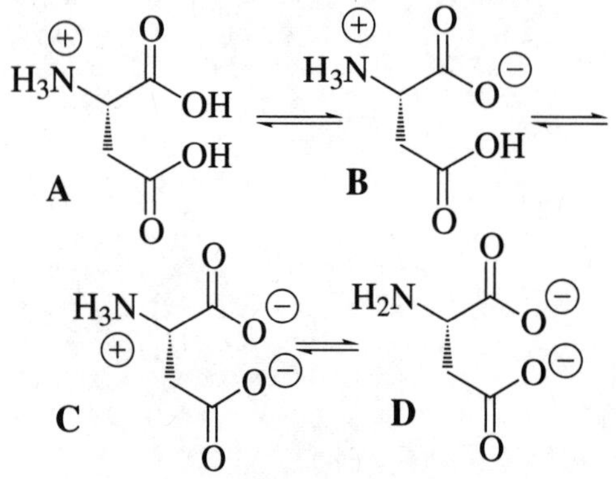

moles of NaOH added

I 100% A; II 50% A, 50% B; III 100% B
IV 50% B 50% C; V 100% C VI 50% C
50% D; VII 100% D

15. **C**   See the solution to question 14.

16. **B**   At pH 7, the carboxyl group is the only group deprotonated to any extent. Since this situation refers to a basic amino acid, it has an extra amino group in its side chain, which is protonated. The diagram to the right shows why the overall charge of this acid is positive. Remember that the cathode is negatively charged and attracts the cation or positively charged species.

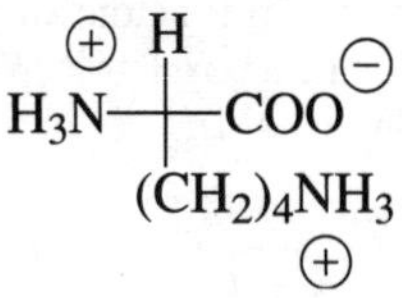

$$\overset{\oplus}{H_3N}\text{---}\overset{H}{|}\text{---}COO^{\ominus}$$
$$(CH_2)_4NH_3$$
$$\underset{\oplus}{}$$

17. **A**   At pH 13.0, more than 50% of the side-chain amino groups are deprotonated. The α-amino and α-carboxyl groups are deprotonated and therefore a net negative charge exists; that is, the amino acid migrates to the positively charged anode.

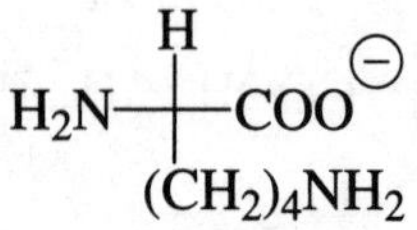

$$H_2N\text{---}\overset{H}{|}\text{---}COO^{\ominus}$$
$$(CH_2)_4NH_2$$

18. **D**   In a strongly acid solution, lysine exists in a fully protonated form and therefore has two positive charges.

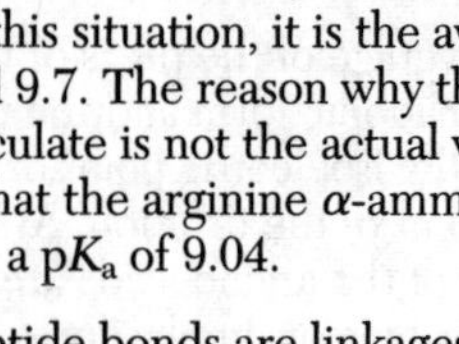

$$\overset{\oplus}{H_3N}\text{---}\overset{H}{|}\text{---}COOH$$
$$(CH_2)_4NH_3$$
$$\underset{\oplus}{}$$

19. **C**   To calculate the isoelectric point, determine the pK_a's of the monocationic and neutral forms, and take the average. In this situation, it is the average of 12.5 and 9.7. The reason why the student's calculate is not the actual value of the pI is that the arginine α-ammonium group has a pK_a of 9.04.

20. **B**   Peptide bonds are linkages between the carbonyl of one amino acid and the amino group of another amino acid, resulting in an amide bond.

21. **C**   The mechanism involved in the formation of peptide bonds is the same as the one involved in the formation of amides. A dehydration occurs and water is released. This backward mechanism is the one shown in the solution to question 22.

22. **D** The acid acts to protonate the oxygen of the carbonyl, allowing water to act as a nucleophile and attack the electrophilic carbon of the carbonyl.

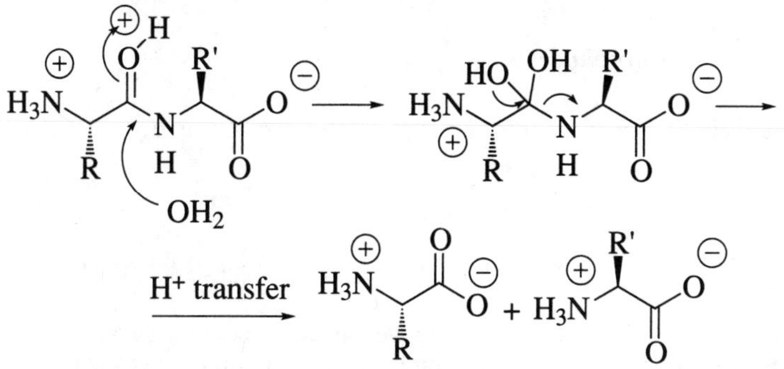

23. **C** See the solution to question 22.

24. **B** See the solution to question 22.

25. **C** Two peptide (amide) bonds are connecting three amino acids, resulting in a tripeptide.

26. **C** When the net peptide is acidic (because of the side chains), the pI values are in the acidic range (< 7).

27. **A** Primary structure refers to the amino acid sequence. The higher forms of amino acid structure refer to the interaction of amino acid groups of peptides. Only the sequence is given; therefore, choice A is correct.

28. **D** $1°$ structure involves the sequence of the peptide chain. The rest all involve $2°$ and $3°$ structures.

29. **D** Tertiary structure involves the coiling or folding of peptide chains stabilized through hydrogen bonds, van der Waals forces, hydrophobic forces, electrostatic forces, and covalent bonds (disulfide linkages). Peptide bonds, however, stabilize the $1°$ structure.

30. **B** Primary structure is the most stable and is the least affected by heat. Secondary and tertiary structures are held together by weaker forces and are therefore more susceptible to degradation by heat. This problem indicates that a monomeric enzyme was degraded and therefore does not have quaternary structure; it consists of only one peptide chain.

31. **D** The noncovalent forces holding the individual strands (making up the collagen) together are disturbed, indicating that the quaternary structure has been disrupted.

Section III: Organic Molecules of Biological Importance

Carbohydrates

1. Which of the following compounds is NOT a monosaccharide?

   A. Ribose
   B. Fructose
   C. Sucrose
   D. Glucose

2. Monosaccharides are classified according to:

   A. the number of carbon atoms in the molecule.
   B. whether they contain an aldehyde or a ketone group.
   C. their configurational relationship to glyceraldehyde.
   D. all of the above.

3. To which groups do glucose and fructose, respectively, belong?

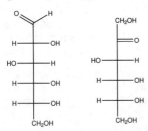

   D-Glucose   D-Fructose

   A. Aldotriose and aldopentose
   B. Ketohexose and aldopentose
   C. Ketohexose and ketohexose
   D. Aldohexose and ketohexose

4. α-D-Glucopyranose is a(n):

   A. hemiacetal.
   B. hemiketal.
   C. acetal.
   D. ketal.

5. Methyl-β-D-glucopyranose is a(n):

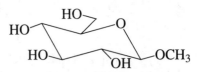

   A. hemiacetal and a reducing sugar.
   B. hemiketal and a nonreducing sugar.
   C. acetal and a nonreducing sugar.
   D. ketal and a reducing sugar.

Questions 6–10 refer to the following molecule:

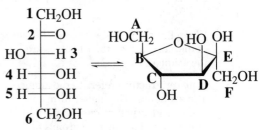

6. This sugar is a(n):

   A. hemiacetal.
   B. hemiketal.
   C. acetal.
   D. ketal.

7. This molecule is a(n):

   A. reducing sugar.
   B. nonreducing sugar.
   C. oxidizing sugar.
   D. all of the above.

8. The preceding cyclic molecule derived from the preceding linear molecule is:

   A. the only possible furanose stereoisomer.
   B. one of two possible furanose enantiomers.
   C. one of two possible furanose diastereomers.
   D. one of eight possible furanose diastereomers.

9. The cyclic ring is formed when:

    A. the hydroxyl group of carbon 6 attacks the carbonyl carbon 2.

    B. the hydroxyl group of carbon 1 attacks the oxygen atom of carbon 5.

    C. the hydroxyl group of carbon 1 attacks the carbon 5.

    D. the hydroxyl group of carbon 5 attack the carbonyl carbon 2.

10. Carbons 1 and 6 in the open-chain form correspond, respectively, to which lettered carbons in the closed ring form?

    A. Carbons A and E

    B. Carbons E and A

    C. Carbons A and F

    D. Carbons F and A

11. Which molecule(s) is (are) NOT commonly found in nature?

    A. L-glyceraldehyde

    B. D-glyceraldehyde

    C. Both are found in nature.

    D. Neither are found in nature.

Questions 12–15 refer to the following information:

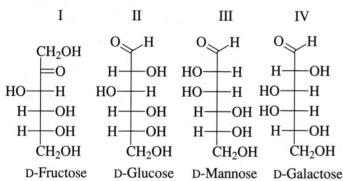

    I — D-Fructose   II — D-Glucose   III — D-Mannose   IV — D-Galactose

12. Which preceding molecule(s) is(are) aldohexose(s)?

    A. I and II

    B. I, II, and III

    C. II, III, and IV

    D. II

13. Which molecule is a C_2 epimer of D-glucose?

    A. I   **C.** IV

    B. III   **D.** II

14. Which molecule differs from D-glucose in configuration at carbon 4 only?

    A. I   **C.** IV

    B. III   **D.** II

15. Which molecules exist as closed cyclic rings in water?

    A. I and II

    B. I, II, and III

    C. II and III

    D. All exist predominantly as cyclic structures.

Using x-ray diffraction studies, a researcher has been able to partially deduce the structure of an unknown monosaccharide. All evidence indicates that it is very similar in structure to D-glucose. Use the following structure to answer questions 16–19.

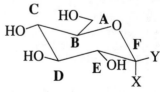

16. Which lettered carbon is known as the anomeric carbon?

    A. C   **C.** E

    B. D   **D.** F

17. Which statement is NOT true?

    A. Epimers have different physical
       properties (e.g., melting point, optical
       properties).
    B. Epimers are diastereomers.
    C. Anomers are enantiomers.
    D. Anomers have different physical
       properties (e.g., melting point, optical
       properties).

18. The preceding molecule is more stable when:

    A. the hydroxyl group at carbon F is at
       position X.
    B. the hydroxyl group at carbon F is at
       position Y.
    C. both molecules are equally stable whether
       X or Y is the hydroxy group.
    D. the compound is not stable with the
       hydroxy group at X or Y.

19. The reasoning behind the answer to question
    18 is that:

    A. intramolecular repulsive forces are at a
       minimum when the hydrogen atom is at
       position X.
    B. intramolecular repulsive forces are at a
       minimum when the hydrogen atom is at
       position Y.
    C. there is no difference in intramolecular
       repulsive forces when the hydroxyl group
       is positioned at X or Y.
    D. There is a difference in the repulsive
       forces, but this difference is small
       compared to the high energy (instability)
       involved with placing a hydroxy group at
       either position.

20. If one adds only pure crystallized α-D-
    glucopyranose to distilled water, allows the
    aqueous solution to sit for 3 hours, and then
    measures the optical properties of the
    solution, the specific rotation value of the
    solution would be:

    A. that of α-D-glucopyranose only.
    B. closer to the value of β-D-glucopyranose.
    C. exactly the average of α and β-D-
       glucopyranose.
    D. closer to the value of α-D-glucopyranose.

**Questions 21–24 are based on the following
disaccharide:**

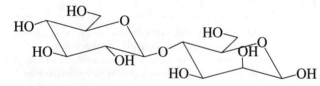

21. What type of covalent bond links these
    monosaccharides?

    A. A 1,4-peptide bond
    B. A 1,4 ether linkage
    C. A 1,4-α-glycosidic linkage
    D. A 1,4-β-glycosidic linkage

22. What type of linkage is synonymous with the
    answer to question 21?

    A. An ether linkage
    B. An acetal linkage
    C. A hemiacetal linkage
    D. An epimeric linkage

23. Hydrolysis of this disaccharide is BEST
    accomplished with:

    A. water alone.
    B. water with acid.
    C. water, Cl_2.
    D. water, CH_3OH.

24. Following hydrolysis and equilibration, the
    resulting products are BEST considered:

    A. anomers.
    B. epimers.
    C. diastereomers.
    D. enantiomers.

SOLUTIONS

Carbohydrates

1. **C** Sucrose (table sugar) is a disaccharide made from the glycosidic linkage of glucose and fructose.

2. **D** The statements are the three main classification criteria for monosaccharides.

3. **D** Glucose has six carbons and an aldehyde group in the straight-chain form and is therefore considered an aldohexose. Fructose has six carbons and a ketone group and is therefore considered a ketohexose.

4. **A** Following are a hemiacetal and glucopyranose. Notice that the monosaccharide is a cyclic hemiacetal. The hydroxy group of carbon 5 attacks the aldehyde at carbon 1 to form the hemiacetal. Remember that the names of the monosaccharide can include the relationship of the ring to furan or pyran.

pyranose

hemiacetal

furan

pyran

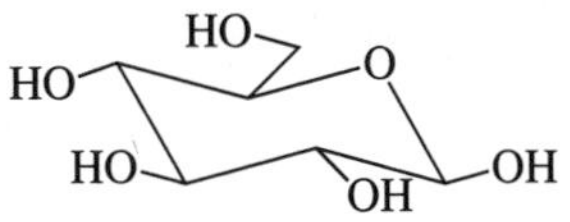

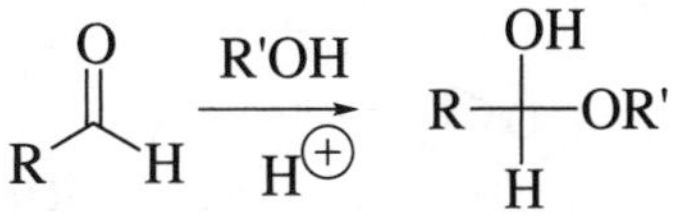

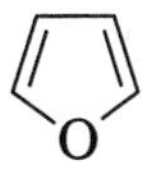

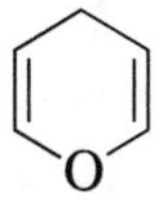

5. **C** Note that the structure for this compound is similar to the structure for an acetal. Sugars with hemiacetal or hemiketal groups are known as reducing sugars because they exist in equilibrium with the open form and the other cyclic anomer. In the open form, the carbonyl is regenerated and gives a positive Tollen's test (involves the reduction of Ag^{2+} to Ag). The equilibrium between cyclic and linear forms does not exist for a cyclic acetal (more difficult to hydrolyze) and never tests positive for the Tollen's test. Therefore, it is called a nonreducing sugar.

6. **B** The cyclic compound has a hemiketal linkage.

7. **A** The cyclic hemiketal shown is a reducing sugar.

8. **C** Two different furanose anomers form because the two different modes of attack by the hydroxy group of carbon 5.

β furanose α furanose

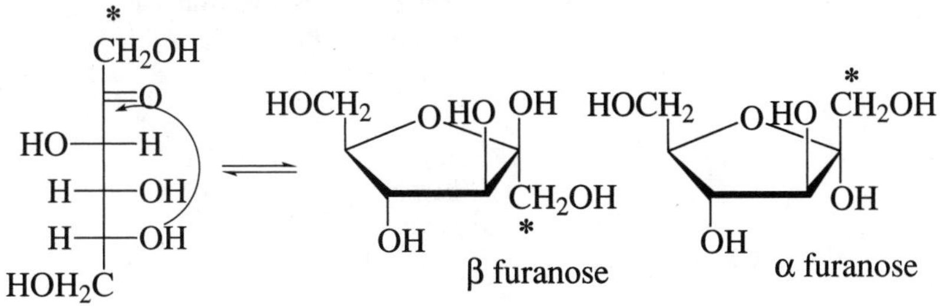

9. **D**   The mechanism is shown in the solution to question 8.

10. **D**   The mechanism in the solution to question 8 makes it possible to trace through the transition from linear to cyclic forms.

11. **A**   Glyceraldehyde is the simplest monosaccharide and is found in two enantiomeric forms, designated D (+: positive optical rotation) or L (−: negative optical rotation). These two compounds serve as configurational standards for all monosaccharides. A monosaccharide (linear form) whose highest numbered chiral carbon (furthest from carbonyl) has the same configuration as D-glyceraldehyde is designated a D sugar. In nature, the D-form of glyceraldehyde is the enantiomer that most commonly occurs.

12. **C**   The structures indicate that they are all hexoses, but only D-fructose is a ketose.

13. **B**   Epimers are diastereomers that differ in configuration only at carbon 2. Note that D-glucose and D-mannose differ in the position of their hydroxyl groups at C-2.

14. **C**   Remember that numbering begins from the end closest to the highest oxidized carbon. The opposite diastereomer is the Fischer diagram with the hydroxy group on the opposite side.

15. **D**   All the structures are hexoses and exist in equilibrium in the cyclic form.

16. **D**   Carbon 1 is considered the anomeric carbon and corresponds to carbon F.

17. **C**   Enantiomers involve the opposite configuration at each chiral center. Epimers and anomers are diastereomers because they differ in the configuration of only one chiral center. Remember that diastereomers have different physical properties.

18. **B**   The hydroxyl group at position Y (equatorial position) results in less steric repulsion between the hydroxy group and the axial hydrogens of the ring.

19. **B**   See the solution for question 18.

20. **B**   This question illustrates mutarotation. Mutarotation is the spontaneous loss of optical at carbon 1. In water, the hemiacetal is hydrolyzed, generating the linear form that can then recyclize to form either the α or β cyclic hemiacetal. The general equilibrium is reestablished over time, and the optical rotation value reflects the fact that the β form is slightly more stable and therefore predominates. Typically, the β form predominates 64:16 over the α anomer.

21. **D**   The β linkage is shown in the diagram that precedes question 21. The linkage is named for the configuration of carbon 1 (sugar on the left). Following is an α-1,4 linkage:

22. **B**   Glycosidic linkages are acetal linkages formed by the loss of water when the hydroxyl group of one monosaccharide attacks the hemiacetal of another monosaccharide.

23. **B**   Similar to the hydrolysis of simple acetals, glycosidic linkages are best hydrolyzed by water and acid or by enzymes.

24. **C**   Note that the two monosaccharides differ in configuration at carbon 2. At hydrolysis, they are epimers (which are diastereomers), but equilibration mutarotation occurs, resulting in four diastereomers (two from each monosaccharide).

Section III: Organic Molecules of Biological Importance

Carboxylic Acid Derivatives: Amides and Esters

1. The correct IUPAC name for the following compound is:

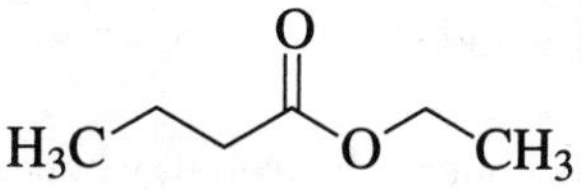

A. ethyl butanoate.
B. isobutylethanoate.
C. 4-ethyl butan-4-oate.
D. 1-ethyl butan-1-oate.

2. The correct IUPAC name for the following compound is:

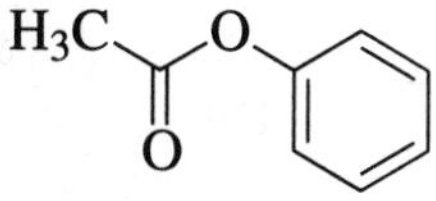

A. ethyl phenyloate.
B. methyl pentanoate.
C. 1-phenyl-1,1-ethanoate.
D. phenyl ethanoate.

3. The correct IUPAC name for the following compound is:

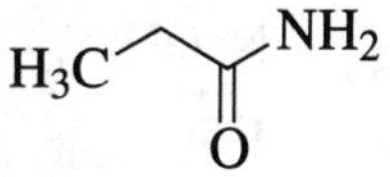

A. propanamine.
B. ethylamine.
C. propanamide.
D. ethylamide.

4. The correct IUPAC name for the following compound is:

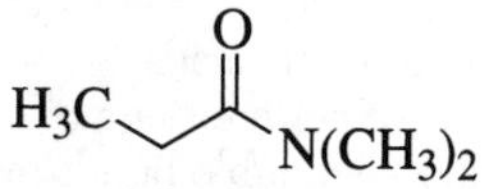

A. 1-dimethyl propanamide.
B. N,N-dimethyl propanamide.
C. N,N-dimethyl propamine.
D. N-dimethyl propanamide.

5. Ethyl methanoate has a boiling point closest to that of:

A. propanol.
B. propanoic acid.
C. propanone.
D. propanamide.

6. Rank the following compounds in order of decreasing boiling point:

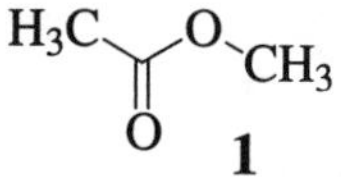

A. 1, 3, 2
B. 3, 1, 2
C. 1, 2, 3
D. 3, 2, 1

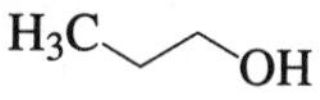

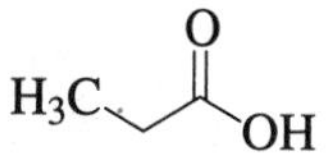

7. Consider a small aliphatic 1° amine, 1° alcohol, ester, carboxylic acid, and amide, all of similar molecular weight. Which one is most likely to have the highest boiling point?

A. The amide
B. The alcohol
C. The ester
D. The carboxylic acid

8. The most stable carboxylic acid derivaties are:

A. acid chlorides.
B. esters.
C. amides.
D. amines.

9. Considering all compounds to be aliphatic and of the same molecular weight, the ones that tend to be the least soluble in water are:

A. 1° amines.
B. carboxylic acids.
C. esters.
D. amides.

10. Esters can be formed by nucleophilic substitution reactions of:

    A. carboxylic acids and alcohols; acid catalyzed.
    B. acid chlorides and alcohols.
    C. both of the above.
    D. neither of the above.

11. Hydrolysis of an ester can be accomplished by:

    A. base-promoted hydrolysis.
    B. H_3O^+ catalyzed hydrolysis.
    C. both of the above.
    D. neither of the above.

Questions 12–14 refer to the following saponification of an ester by sodium hydroxide:

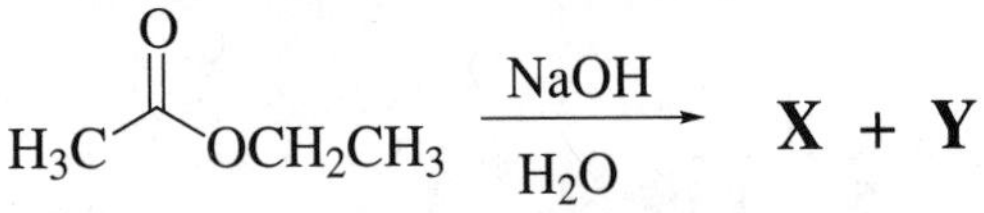

$$\xrightarrow[H_2O]{NaOH} X + Y$$

12. The Products **X** and **Y** are:

    A. $CH_3OH + CH_3CH_2OH$.
    B. $CH_3COOH + CH_3CH_2O^-Na^+$.
    C. $CH_3COO^- Na^+ + CH_3CH_2OH$.
    D. none of the above.

13. This reaction occurred by:

    A. a nucleophilic attack by base at the acyl carbon.
    B. nucleophilic attack by base at the alkyl carbon of the alcohol portion of the ester.
    C. nucleophilic attack by base at the α-carbon.
    D. none of the above.

14. The alcohol portion of the ester usually:

    A. inverts configuration.
    B. retains configuration.
    C. undergoes full racemization.
    D. changes configuration of the anomeric carbon.

15. In the following ester hydrolysis reaction, in which product does labeled ^{18}O appear?

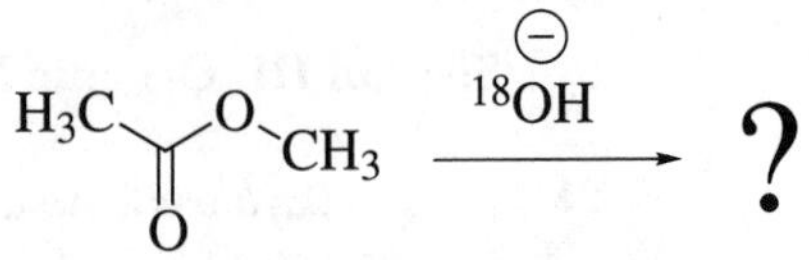

    A. The alcohol
    B. Carboxylate ion
    C. Carboxylic acid
    D. Hydroxide ion

16. Amides undergo hydrolysis when they are heated in:

    A. aqueous acid.
    B. aqueous base.
    C. both of the above.
    D. neither of the above.

17. In the following reaction, **Z** must be:

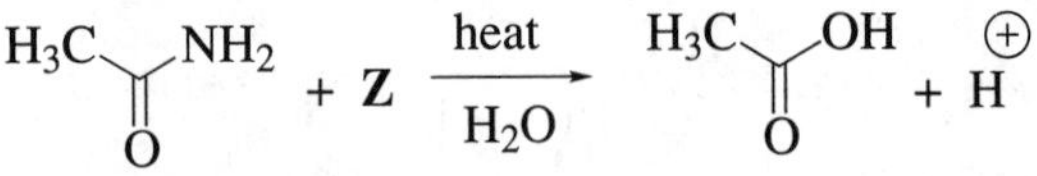

    A. OH^-.
    B. H_3O^+.
    C. H_2.
    D. CH_3OH.

18. In the following reaction, **Y** must be:

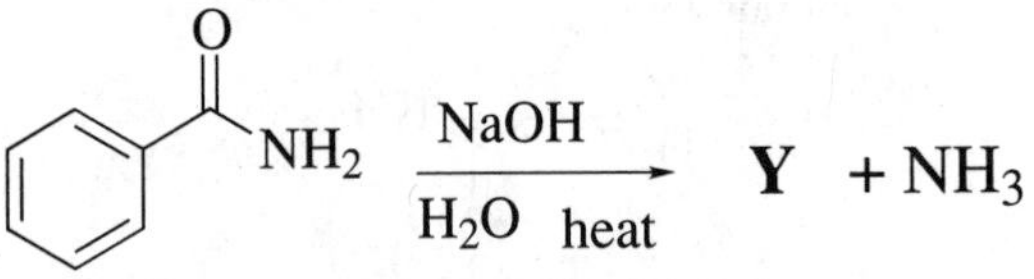

    A. benzoic acid.
    B. 3-hydroxybenzoic acid.
    C. benzene.
    D. sodium benzoate.

19. Which statement is NOT true about oils?

    A. Oils possess high percentages of unsaturated fatty acids.
    B. Oils have lower melting points than fats.
    C. Oils have *cis*-double bonds in their fatty acids.
    D. Oils usually consist of unbranched carboxylic acids with odd numbers of carbon atoms.

20. Triglycerides are esters of long-chain fatty acids and glycerol. At hydrolysis with NaOH, which product do triglycerides produce?

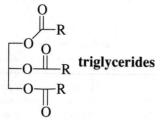

triglycerides

A. All salts
B. Salt and carboxylic acids
C. Glycerol and sodium carboxylates
D. Alcohol and carboxylic acids

21. Soaps are:

I. long-chain esters.
II. long-chain fatty acids.
III. sodium salts of long-chain carboxylic acids.
IV. both polar and nonpolar molecules.
V. emulsifiers.

A. I, IV, and V
B. II and V
C. III, IV, and V
D. II, IV, and V

22. Which BEST accounts for the fact that soap dissolves in water?

A. H-bonding occurs on both ends of soap molecules.
B. Soap molecules are exclusively polar and form strong bonds with water.
C. Strong van der Waals forces attract soap and water molecules.
D. Micelles form

SOLUTIONS

Amides and Esters

1. **A** To name an ester, identify the longest carbon chain that contains the carbonyl group. In this compound, it is butane. Drop the "e" and add "oate" to the longest carbon chain to get butanoate. The carbon group attached to the carboxyl OH group is named an alkyl group. In this compound, the name is ethyl. Therefore, the compound is ethyl butanoate.

2. **D** The longest carbon chain that contains the carbonyl group is a two-carbon chain. The alkyl group attached to the carboxyl group OH group is a phenyl group. Therefore, the compound is named phenyl ethanoate. It is not necessary to give number locations for attachment of the alkyl group.

3. **C** To name an amide, identify the carbon chain containing the carbonyl group that is attached to the nitrogen. In this compound, it is a three-carbon chain. The compound is therefore known as propanamide. Note that the "e" is dropped from the hydrocarbon name and is replaced by the suffix "amide."

4. **B** Because this compound has a carbonyl group attached to a nitrogen, it is an amide. The carbon chain containing the carbonyl group is a three-carbon chain and so similar to question 3, that its base name is propanamide. The additional substituents are methyl groups, which are directly attached to the nitrogen. They are listed with an "N" to indicate that they are attached to the nitrogen. Since two methyl groups are attached to the N, two N's are listed.

5. **C** Ethyl methanoate is an ester containing three-carbons. Esters have boiling points closest to that of ketones and aldehydes of similar molecular weight because all three of these compounds cannot form H-bonds with themselves. On the other hand, propanol, propanoic acid, and propanamide all form strong H-bonds with themselves because they all possess good H-bond donors and acceptors.

6. **D** First note that all three of these compounds have similar molecular weight because they all contain three carbons and one or two oxygens. The primary factor in differentiating their boiling points must then be hydrogen bonding ability. The carboxylic acid (3) has the greatest H-bonding ability because it contains the most acidic proton and a strong H-bond acceptor (carbonyl). The alcohol is next followed by the ester (does not contain a good H-bond donor).

7. **A** Amides can form very strong H-bonds because of the ability of the N to donate electrons to the carbonyl (more so than OH of a carboxylic acid) and give the O considerable negative charge, thereby enhancing the H-bond acceptor capability of the molecule. Considering compounds of similar molecular weight and carbon chain structure, the general trend is (highest to lowest boiling point): amides, carboxylic acids, alcohols, amines, aldehydes/ketones/esters, alkynes, alkenes, and alkanes.

8. **C** Amides are the most stable of the carboxylic acid derivatives because they have the most stable resonance forms:

9. **C** Choices A, B, and D can form strong H-bonds with water. Although the esters can form H-bonds with water, they do not contain H's which can be that as H-bond donors. Therefore, the H-bonds that esters form with water are considerably weaker than the H-bonds formed by the other choices.

10. **C** The following reactions show the mechanisms of the two reactions:

In acid:

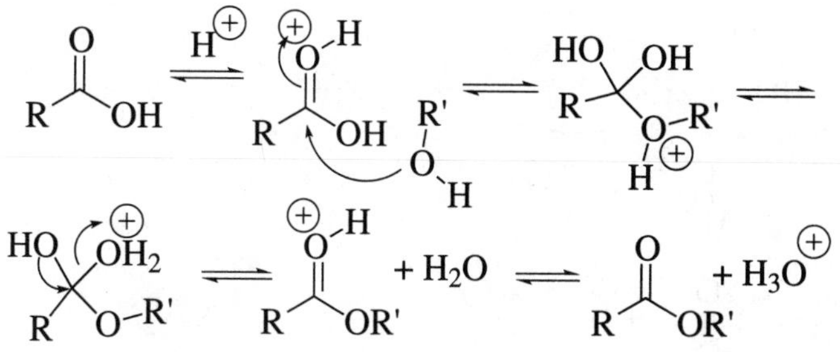

Acid chloride with an alcohol:

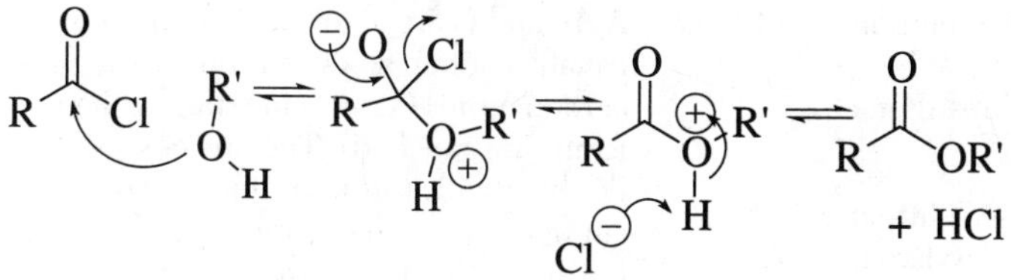

11. **C** The following reactions show how both base or acid hydrolysis of esters can occur.

In base:

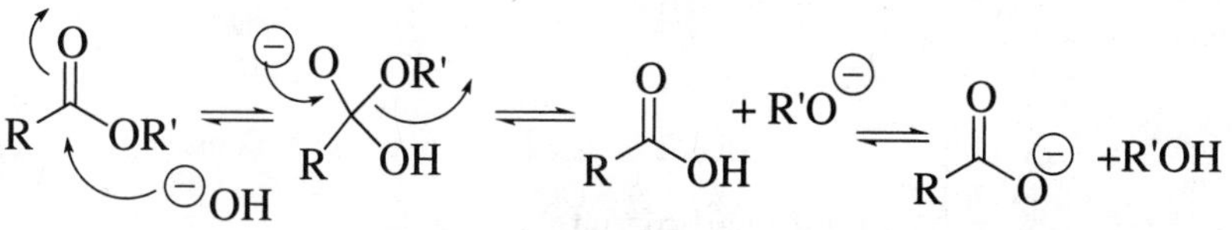

The reverse of the mechanism listed for the solution to question 10 is the forward mechanism for acidic ester hydrolysis.

12. **C** Base hydrolysis of esters (saponification) produces alcohols and salts of carboxylic acids (sodium carboxylates). The alkoxide leaving group is more basic than the carboxylate; therefore, the carboxylic acid is deprotonated to give the carboxylate and the alcohol.

13. **A** The preceding mechanism for base hydrolysis of esters (solution to question 11) shows that the acyl carbon is attacked by the base. It does this via a nucleophilic substitution reaction.

14. **B** In S_N2 reactions, an overall inversion of configuration of the molecule follows substitution. However, with an ester with a chiral alcohol moiety, the alcohol retains its configuration when it is released.

15. **B** The mechanism for the solution to question 11 depicts what occurs for the labeling experiment and indicates that the label will appear on the carboxylate.

16. C The hydrolysis of amides occurs in both acid and base. Basic hydrolysis is similar to that of esters shown in the solution for question 11. Acidic hydrolysis is shown next:

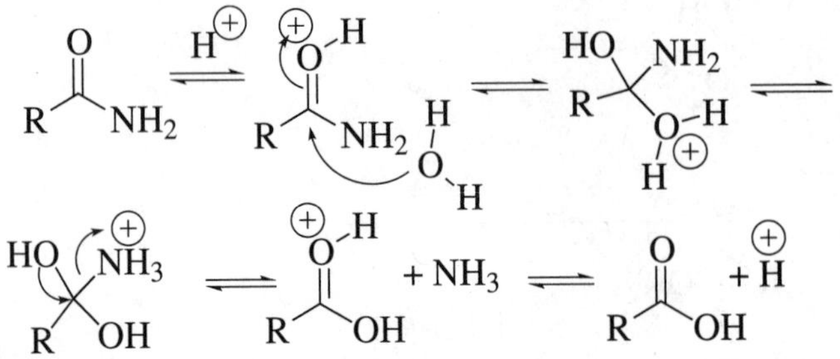

17. B The mechanism outlined for the solution to question 16 indicates that Z must be an acid. Basic hydrolysis would result in a carboxylate.

18. D The mechanism outlined for the solution to question 11 shows that a carboxylate is formed and, here since NaOH is used, a sodium carboxylate salt results.

19. D A, B, and C are all true. Fats and oils usually contain an even number of carbon atoms (12–24) in straight chain form (unbranched). The oils possess double bonds (unsaturation) in *cis* configuration. Therefore, they have lower melting points than saturated fats.

20. C Because triglycerides are just esters of alcohols (glycerol) and fatty acids (carboxylic acids), their products are most likely the alcohol and sodium carboxylate salts.

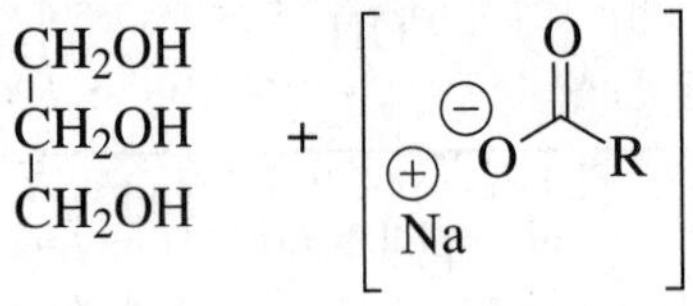

21. C The sodium carboxylate salts of long-chain fatty acids are called soaps. They have a polar end and a long nonpolar region. When a particular concentration is reached, they form micelles. These structures are spherical, with the hydrophobic chains comprising the center and the polar carboxylates along the outside, thereby making these structures water-soluble. They are known as emulsifiers because they are capable of holding together in one phase two mutually immiscible substances. The hydrophobic component of a mixture is absorbed into the micelle, and the micelle remains water-soluble because of the polar carboxylate groups.

22. D Only the polar end of the soap molecule can form H-bonds. The hydrocarbon is hydrophobic, and it is not energetically favorable for it to associate with water. Van der Waals forces are always weak forces. Therefore, A, B, and C are incorrect. Micelles (discussed in the solution to 21) "hide" the hydrophobic regions of the fatty acid from the solvent while the polar ends interact with the polar solvent, making the whole structure soluble.

186